Paleobiology of the Neoproterozoic Svanbergfjellet Formation, Spitsbergen

NICHOLAS J. BUTTERFIELD, ANDREW H. KNOLL AND KEENE SWETT

Butterfield, N.J., Knoll, A.H. & Swett, K. 1994 07 15: Paleobiology of the Neoproterozoic Svanbergfjellet Formation, Spitsbergen. *Fossils and Strata*, No. 34, pp. 1–84. Oslo. ISSN 0300-9491. ISBN 82-00-37649-4.

A fossil *Lagerstätte* from the 700–750 Ma old Svanbergfjellet Formation of northeastern Spitsbergen offers a substantially enhanced view of late Proterozoic paleobiology. Fossils occur primarily as organic-walled compressions in shales and permineralizations in chert; secondary modes of preservation include bedding-plane imprints and mineral replacements in apatite and goethite(?). The depositional setting of all fossiliferous horizons is broadly peritidal with highest taxonomic diversity occurring in shallow subtidal settings; the details of included fossil assemblages contribute to improved paleoecological resolution. The often distinct constituents of shale- and chert-hosted fossil assemblages appear to be a product of both paleoenvironment and fundamentally dissimilar taphonomic pathways, such that only forms with inferred wide ecological tolerance appear in both. Consideration of taphonomic processes also provides a variety of useful taxonomic insights, on the one hand permitting some resolution of so-called wastebasket taxa, such as *Chuaria*, and on the other acknowledging the taxonomic disparity that can occur in simple forms like *Siphonophycus* and *Oscillatoriopsis*. True multicellular (including coenocytic) eukaryotes are a conspicuous component of the Svanbergfjellet assemblage: of eight distinct taxa, one can be identified as a coenobial/colonial chlorococcalean and three as filamentous siphonocladaleans (Chlorophyta). Other forms are problematic, but several show significant cell, or possibly tissue, differentiation. A review of Proterozoic multicellular organisms reveals that a coenocytic grade of organization was common among early metaphytes and supports the view that cellularity is a derived condition in many 'multicellular' lineages. Nineteen acritarch taxa are preserved in the Svanbergfjellet sediments. Ten of these show a readily identifiable ornamentation and contribute significantly to Neoproterozoic biostratigraphy; a world-wide and exclusively Late Riphean distribution of the acanthomorph *Trachyhystrichosphaera aimika* identifies it as a particularly valuable index fossil. The Svanbergfjellet fossil assemblage preserves a total of 63 distinct forms, of which 56 are treated taxonomically. As much as possible, principles of 'natural' taxonomy are applied, such that taphonomic and ontogenetic variants are declined separate taxonomic status. Major taxonomic revisions are offered for the acritarchs *Trachyhystrichosphaera* and *Chuaria* as well as for the prokaryotic-grade filaments: *Cephalonyx, Cyanonema, Oscillatoriopsis, Palaeolyngbya, Rugosoopsis, Siphonophycus, Tortunema,* and *Veteronostocale*. Newly erected taxa include 7 new genera: *Palaeastrum, Proterocladus, Pseudotawuia, Valkyria, Cerebrosphaera, Osculosphaera* and *Pseudodendron*; 14 new species in 12 genera: *Palaeastrum dyptocranum, Proterocladus major, Proterocladus minor, Proterocladus hermannae, Pseudotawuia birenifera, Valkyria borealis, Cerebrosphaera buickii, Osculosphaera hyalina, Pseudodendron anteridium, Dictyotidium fullerene, Germinosphaera jankauskasii, Trachyhystrichosphaera polaris, Siphonophycus thulenema* and *Digitus adumbratus*; and 7 new combinations: *Leiosphaeridia wimanii, Eoentophysalis croxfordii, Cephalonyx geminatus, Oscillatoriopsis amadeus, Siphonophycus typicum, Siphonophycus solidum* and *Tortunema Wernadskii*. □ *Proterozoic, Neoproterozoic, microfossils, macrofossils, problematica, acritarchs, biostratigraphy, taphonomy, Svalbard, Spitsbergen, multicellularity, metaphytes, algae, cyanobacteria, microbial mats, Chlorophyta, taxonomy.*

Nicholas J. Butterfield and Andrew H. Knoll, Department of Organismal and Evolutionary Biology, Harvard University, Cambridge, Massachusetts, USA. 02138 (NJB present address: Department of Earth Sciences, University of Cambridge, Cambridge, UK CB2 3EQ); Keene Swett, Department of Geology, University of Iowa, Iowa City, Iowa, USA. 52242; received 1993 02 15.

Contents

Introduction

Instances of exceptional fossil preservation are becoming increasingly appreciated as the best available measures of paleodiversity (Conway Morris 1986). Accordingly, it is the rare fossil *Lagerstätten* such as the Burgess Shale or the Solnhofen Limestone that have come to serve as the principal reference points for reconstructing evolutionary history. The later Proterozoic is among those intervals most calling for comparable *Lagerstätte*-type documentation as it holds the immediate forebears and, potentially, the explication of the Ediacaran–Cambrian radiation of large organisms (Knoll 1991). Exceptional fossil preservation is known in rocks of this age, but much of it is limited to prokaryotic microbial-mat assemblages in relatively restricted carbonate facies (e.g., Schopf 1968; Knoll 1982); shale-hosted *Lagerstätten* are typically more diverse (e.g., Timofeev & Hermann 1979; Hermann 1981a, b) and can also be taxonomically distinct. A reasonably representative accounting of Neoproterozoic paleobiology thus requires superior fossil preservation in various taphonomic modes and from diverse paleoenvironments.

Sediments of the 700–800 Ma old Akademikerbreen Group, northeastern Spitsbergen (Fig. 1), have proven a rich source of silicified microfossils and related paleobiological data, particularly in the Draken Conglomerate (Knoll 1982; Knoll *et al.* 1991) and overlying Blacklundtoppen formations (Knoll *et al.* 1989). Shale-hosted fossil assemblages are also reported from these units, but truly exceptional preservation in siliciclastic facies appears to be limited to the immediately subjacent Svanbergfjellet Formation (Butterfield *et al.* 1988). As the Svanbergfjellet also preserves a diverse silicified biota and various mineralized forms, it offers an unusually complete view of the late Proterozoic biosphere as it approached the Ediacaran and Cambrian radiations. The Svanbergfjellet fossil *Lagerstätte* is here described with particular reference to paleoenvironmental distribution, taphonomic processes, multicellular and unicellular eukaryotes, and various taxonomic revisions.

Materials and methods

Paleontological investigation of the Svanbergfjellet Formation was directed primarily at organic-rich (i.e. dark) chert nodules in carbonate facies and fine-grained siliciclastic rocks. Cherts were examined in petrographic thin sections cut perpendicular to bedding; of 39 samples, 22 proved to be fossiliferous, half of these exceptionally so (Fig. 2). Careful acid (HF) maceration of fossiliferous chert samples yielded no intact fossil material.

A standard palynological maceration of the Svanbergfjellet shales produced a number of taxonomically depauperate fossil assemblages comprised largely of fragmentary material. Further investigation by way of bedding-parallel thin section revealed that much of this fragmentation was a product of the extraction procedure and that several shale horizons preserved abundant and diverse fossils. Indeed, the conventional palynological treatment of most Proterozoic shales may be responsible for a significant bias in the early paleontological record through its preferential recovery of simple and relatively robust fossils. Study in thin section also preserves biologically and paleoenvironmentally significant details of fossil associations, orientation, and bedding-plane distribution (e.g., Butterfield & Chandler 1992). Such being the case, all of the Svanbergfjellet shales were initially examined in petrographic thin sections cut parallel to bedding.

The principal advantage to maceration techniques lies in the ability to sample significant volumes of material. A simple, low-manipulation maceration procedure was therefore developed to enhance the recovery of rare fossil taxa, especially those too large or too delicate to be extracted by conventional techniques. The procedure involved submersing small (ca. 1 cm³) uncrushed subsamples of fossiliferous shale in concentrated hydrofluoric acid (48% HF) where they were allowed to disaggregate with minimal agitation. Following two or three rinses in distilled water the preparations were examined for fossils with a binocular stereoscope using both transmitted and reflected light. Isolated fossils were collected by pipette, passed through two baths of distilled water and transferred, always suspended in water, to

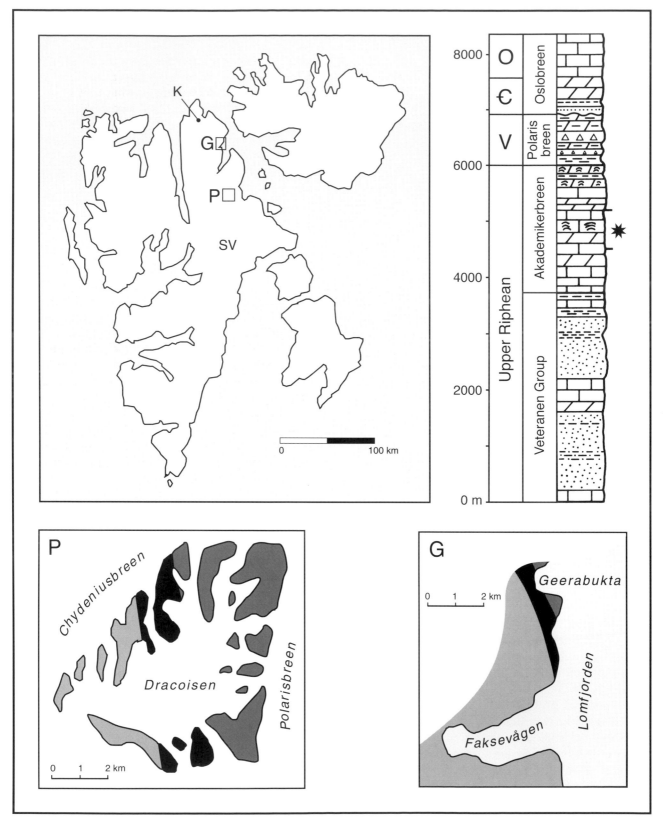

Fig. 1. Map of Spitsbergen and neighboring islands, showing the principal outcrop localities of Svanbergfjellet Formation: G = Geerabukta (79°35'33"N, 17°44'E); P = Polarisbreen (79°10'N, 18°12'E); SV = Svanbergfjellet (78°41'30"N, 18°14'E); K = Kluftdalen (79°50'27"N, 17°E). Black areas in the enlargements G and P indicate Svanbergfjellet Formation outcrop (dark grey areas of outcrop indicate younger strata; lighter grey younger; outcrop at Polarisbreen occurs as a series of nunataks). The stratigraphic column represents the late Proterozoic (Lomfjorden Supergroup) and early Paleozoic sedimentary successions in northeastern Spitsbergen; the Svanbergfjellet Formation is marked with a star.

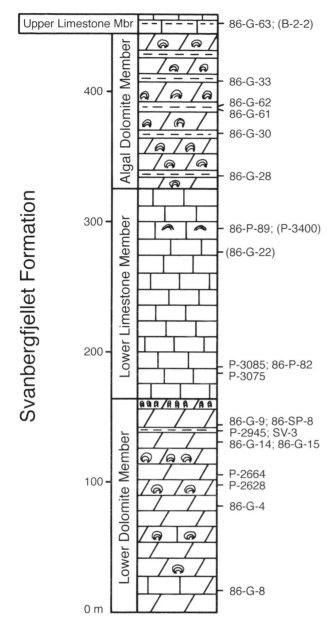

Fig. 2. Stratigraphic column of the Svanbergfjellet Formation as measured at Geerabukta (G), showing the positions of significantly fossiliferous horizons. The level of samples from other localities ('P', 'SP', 'SV', 'B') are inferred. Sample numbers in parentheses represent fossiliferous units not figured in the present work.

microscope-slide cover-slips. Removal of the water droplet results in a fossil adhering to the glass surface where it can then be prepared for light or scanning electron microscopy. The much improved photogenicity of these isolated fossils further encouraged the extraction of several specimens initially identified in thin section (e.g., Figs. 16A, 17C, 19C, 25E).

With few exceptions (e.g., *Proterocladus hermannae* n.sp., *Digitus adumbratus* n.sp.), all taxa recovered by acid maceration were also observed in thin section, thus confirming their

syngenicity with surrounding sediments. The isolated fossil material is additionally distinguished from modern contaminants by its two-dimensionality and a distinct graphitic surface sheen derived from the low-grade burial metamorphism. On average, the particulate organic material is light to medium brown under transmitted white light.

Of 17 sampled shale horizons, 8 contained fossils (Fig. 2). Two of these, P-2945 (Lower Dolomite Member) and 86-G-62 (Algal Dolomite Member), revealed exceptionally well preserved and diverse fossil assemblages, together supplying a majority of the Svanbergfjellet fossil diversity. These green to grey shales are fissile, finely laminated and extremely fine-grained, with most sediment grains measuring less than 1 μm in maximum dimension (Fig. 3E). Compositionally they are almost entirely siliciclastic (no carbonate) and contain less than 1.5% total organic carbon (TOC).

Geological setting

The Svanbergfjellet Formation is a distinctive succession of limestones, dolomites, and subordinate siliciclastic facies within the Neoproterozoic Akademikerbreen Group, northeastern Spitsbergen (Fig. 1; Wilson 1961; Knoll & Swett 1990). Stratigraphically, it lies some 2000 m below rocks of Early Cambrian age and ca. 1000 m below a late Proterozoic (Varanger) tillite. Radiometric dates place few constraints on the age of the Svanbergfjellet Formation, but microfossils in the formation (and in the overlying Draken Conglomerate; Knoll *et al.* 1991) indicate a younger Late Riphean age, ca. 700–800 Ma. Biostratigraphic and chemostratigraphic ($\delta^{13}C$ and $^{87}Sr/^{86}Sr$; Knoll *et al.* 1986; Derry *et al.* 1989; Asmerom *et al.* 1991) correlation with the better dated Shaler Group of northwestern Canada (minimally 723±4 Ma old; Heaman *et al.* 1992) corroborate and potentially constrain this estimate.

The Svanbergfjellet Formation is nearly 600 m thick in its southernmost exposures at Svanbergfjellet nunatak, but thins to an estimated 100 m at Kluftdalen, ca. 150 km to the north (Fig. 1; Wilson 1961). Despite the apparent thickness variation, the same four members can be distinguished throughout the outcrop area (Fig. 2).

The basal Lower Dolomite Member lies conformably above the mixed limestones and dolomites of the upper Grusdievbreen Formation. This member, ca. 150 m thick throughout its area of exposure (save for Kluftdalen), consists predominantly of stratiform microbialites and associated intraclastic grainstones, dolomicrites, and scattered low domal or digitate stromatolitic bioherms (Fig. 3C, D, F). Most carbonates are dolomite, although subordinate thicknesses of limestones occur; a 4–8 meter sequence of quartz arenite and shale near the top of the member constitutes the only significant incursion of siliciclastic rocks. At Polarisbreen this latter unit contains one of the two principal shale-hosted fossil assemblages in the formation (sample P-2945), while correlative shales at Svanbergfjellet include centime-

Fig. 3. □A. Outcrop of the Algal Dolomite Member. The massive units are stromatolitic dolomite, separated by fine-grained siliciclastic rocks. □B. *Minjaria* stromatolites from the top of the Lower Dolomite Member; scale bar equals 6.2 cm. □C, D. Silicified intraclastic grainstone from the Lower Dolomite Member; Sample 86-G-8-2a; the large clast in C and a majority of the smaller clasts in both C and D are of massive, uncompressed, and clastic-free microbial mat. The large clast in D is of the clastic-rich laminated mat type; note the incorporation of various intraclasts, including shards of clastic-free mat; scale bar in B equals

ter- to decimeter-sized apatite nodules that preserve vertically oriented filamentous structures. Silicified intraclastic
carbonates of the Lower Dolomite Member preserve abundant microfossils. Fenestrae and erosional surfaces up to
several centimeters deep document a frequent subaerial exposure of Lower Dolomite carbonates (Knoll & Swett 1990),
and the succession is interpreted as a tidal-flat–lagoonal
complex, not dissimilar from that described for the overlying
Draken Conglomerate (Knoll *et al.* 1991). The member is
capped by a 5–8 m thick *Minjaria* biostrome (Fig. 3B) that
can be traced from Svanbergfjellet north to at least Geerabukta.

The succeeding Lower Limestone Member is a ca. 150 m
thick transgressive sequence. It consists predominantly of
dark grey to black, laminated to decimeter-scale beds of
calcilutite that commonly contain microspar-filled syneresis
cracks. Microbially laminated carbonates are rare within this
member, as is chert. However, there are at least two thin
horizons of silicified stratiform stromatolites, both of which
contain microfossils. Most Lower Limestone deposition
took place below fair-weather wave base, following the broad
trend for deeper-water carbonates in the late Proterozoic to
be limestone rather than dolomite (Knoll & Swett 1990).

The most distinctive unit of the Svanbergfjellet Formation
is the Algal Dolomite Member, which consists of orange-
weathering stromatolitic dolomites interbedded with green
to black (rarely red) siliciclastic mudstones (Fig. 3A). Much
of the thickness variation recorded for the formation as a
whole is accounted for by this member, which varies from
250 m in the south to just 50 m at Kluftdalen; at Geerabukta,
where many of the fossils reported here were collected, the
member is 100 m thick. Algal Dolomite Member stromatolites consist of inzeriform and tungussiform columns that
form domes or of extensive biostromes with billowy surfaces;
as the member's name implies, all carbonates are dolomitic.
Despite the conspicuous nature of the stromatolites in outcrop, mudstones (and rare quartzose sandstones) constitute
about half of the member's thickness. The mudstones drape
stromatolites forming sharp lithological contacts (Fig. 3A)
and constitute the principal locus of microfossil preservation
in the formation (sample 86-G-62). Deposition of the Algal
Dolomite Member appears to have taken place largely in
quiet subtidal environments, although displaced blocks
within several biostrome horizons indicate that strong
storms occasionally affected the bottom.

2.5 mm. □E. Thin section, perpendicular to bedding, of shale sample 86-
G-62 (Algal Dolomite Member); note the very fine grain size and undisrupted lamination; scale bar in B equals 120 μm. □F. Silicified columnar
stromatolite and intraclastic grainstone; Lower Dolomite Member; Sample
86-G-15-2A; scale bar in B equals 15 mm. □G. *Polybessurus bipartitus*
preserved in large dolomite crystals with pronounced cleavage; Lower
Limestone Member; Sample 86-P-89 (S-64-0); scale bar in B equals
120 μm.

The Upper Limestone Member comprises the uppermost
unit of the formation. Like the underlying Algal Dolomite
Member, this succession thins markedly from more than
100 m in the south to a few meters at Kluftdalen. Dark grey to
black limestones, much like those of the Lower Limestone
Member, predominate. This upper unit is distinguished by
its greater proportion of dolomitic grainstones and carbonaceous shales, which increase in abundance toward the top of
the member. A regionally extensive bench of quartz arenite
marks the top of the formation (although not at Geerabukta); Svanbergfjellet rocks are overlain abruptly by the
intraclastic dolomitic grainstones of the Draken Conglomerate.

Paleoenvironments

Overall, the Svanbergfjellet Formation records a variety of
coastal marine depositional environments ranging from
supratidal to subtidal below wave base. Fossil assemblages
document aspects of biological activity over much of this
habitat range. Indeed, paleontological data may often provide as much or more resolution as sedimentary fabrics in
determining paleoenvironments (e.g., Knoll *et al.* 1991).
This is especially true of shale facies, which commonly lack
diagnostic sedimentary structures (Butterfield & Chandler
1992).

In carbonate facies most Svanbergfjellet microfossils are
found within silicified flake conglomerates of ripped-up
microbial mat. This microbialite grainstone was deposited in
lower intertidal to shallow subtidal environments, as indicated by the minimal clast transport (clasts are typically
poorly sorted, angular and of a high aspect ratio; Fig. 3C–D)
and the co-occurrence of *in situ* columnar stromatolites with
several centimeters of synoptic relief (Fig. 3F). Paleontologically, the setting is represented by the large ornamented
acritarchs *Cymatiosphaeroides* and *Trachyhystrichosphaera*,
which occupy the interstitial spaces of grainstones and often
appear to have acted as individual clastic particles. Some
interstices further show evidence of microbial recoloniza-
tion, thus recording some degree of post-depositional biological activity: geopetally oriented populations of *Sphaerophycus* in the matrix of sample 86-SP-8 (Fig. 20L–T), and
interstitial filaments in sample 86-G-15 (Fig. 23B–D) were
almost assuredly photosynthetic and point to relatively extended periods of quiescence well within the photic zone. In
both instances this secondary growth would have contributed to the final stabilization of the sediments.

The sedimentary and paleontological fabrics *within* the
constituent clasts of the microbialite grainstones further
reveal their diverse, if broadly peritidal, provenance. Apart
from non-fossiliferous clasts of (or after) carbonate microspar, two clast types dominate most of the grainstones. The
most common is constructed of densely interwoven mats of
Siphonophycus typicum n.comb. with interspersed populations of *Myxococcoides* (Fig. 3C). These clasts are entirely free

of mineral grains, show no evidence of sedimentary compaction, and have only vague lamination; both the sedimentological and paleoecological evidence point to a source in the high-intertidal zone (cf. Knoll *et al.* 1991). The other principal clast type is characterized by a distinctive fabric of crinkly sapropelic laminae that incorporate substantial amounts of clastic material, including flakes of the dense, non-laminated microbial-mat type (Fig. 3D). Filamentous and other microbial-mat fossils are uncommon as primary constituents, the most conspicuous fossils being the large ornamented acritarchs *Trachyhystrichosphaera* and *Cymatiosphaeroides.* Given the common occurrence of these same acritarchs in the grainstone matrix, the paleontological signature supports the textural evidence for a shallow subtidal derivation of these laminated clasts.

Other paleontologically distinct but less common microbialite intraclasts are suggestive of various micro-environments within the intertidal zone. Some are constructed entirely of dense colonial populations of spheroidal microfossils such as *Eoentophysalis croxfordii* n.comb. (Fig. 20A) or *Eoentophysalis belcherensis* (Fig. 20D) and compare closely with modern intertidal–supratidal entophysalid mats (cf. Golubic & Hofmann 1976). A somewhat different setting is suggested in sample 86-SP-8, where most clasts are dominated by *Sphaerophycus parvum.* In the Lower Limestone Member, two chert samples (86-P-89 and P-3400) contain the intertidal cyanobacterium *Polybessurus bipartitus* (Green *et al.* 1987; Knoll *et al.* 1991), although as *in situ* crusts rather than in redeposited intraclasts. The presence of shallow-water *Polybessurus,* along with fenestrae infilled with botryoidal and concentrically layered cements, identifies the localized shallowing of this otherwise deeper-water unit. Other fossiliferous cherts in the Lower Limestone Member also represent *in situ* sedimentation; however, their even lamination (sample P-3075) or exclusively planktic fossil assemblages (sample P-3085) point to fully subtidal conditions.

All of the fossiliferous shales in the Svanbergfjellet Formation were deposited in relatively shallow water as evinced by their close proximity to, and intercalation with, stromatolitic carbonates in consistently peritidal sequences. Moreover, a non-random distribution of microfossils on bedding planes and abundant laterally extensive microbial mats indicates a significant proportion of benthic, probably photosynthetic organisms (Butterfield & Chandler 1992). Even so, the distinct distributions of fossils in the two principal fossil-bearing shales suggest important differences in paleoenvironment that are not reflected sedimentologically. Where sample P-2945 contains an abundant but taxonomically depauperate assemblage of spheroidal and ornamented acritarchs (including *Trachyhystrichosphaera*), the fossils in sample 86-G-62 are highly diverse, showing little dominance of any particular taxa. Significantly, the size-frequency distributions of planktic leiosphaerids in the two units are markedly dissimilar (Fig. 4). The limited development of filamentous microbial mats or other benthos in sample P-2945

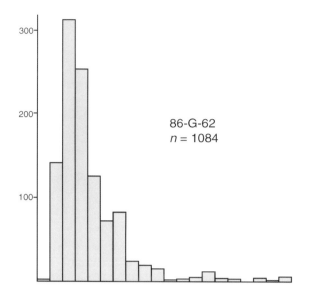

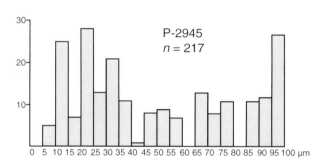

Fig. 4. Size-frequency distribution of leiosphaerid acritarchs less than 100 μm in diameter, from the two principal fossiliferous shales: 86-G-62 (Algal Dolomite Member) and P-2945 (Lower Dolomite Member). Counts are from bedding-parallel thin sections (9 from 86-G-62; 3 from P-2945) and include all apparently planktic (i.e. non-clustered) specimens. Note the differing vertical scales of the two histograms.

suggests that these distributional differences derive from its more distal paleoenvironment relative to that of 86-G-62 (zone 3 *vs.* zone 2 in the scheme of Butterfield & Chandler 1992).

In Proterozoic successions, shale and silicified carbonate and shale facies represent more or less distinct paleoenvironments; however, they broadly intersect in shallow subtidal settings (Knoll 1984; Butterfield & Chandler 1992). Fossil taxa common to both subtidal cherts and shales in the Svanbergfjellet Formation include the acanthomorphic acritarch *Trachyhystrichosphaera,* some leiosphaerids, and a variety of filaments – *Palaeolyngbya, Rugosoopsis* and, especially, *Siphonophycus,* which occurs in both contexts as extensive LPP-type microbial mats. Furthermore, species of *Obruchevella, Veteronostocale* and *Brachypleganon* that were originally described from silicified carbonates are here preserved in shale. Taphonomic selectivity of the two systems will account for some of the remaining differences; however, much of it surely derives from basic ecological differences between clear-water (carbonate) and muddy (shale) paleo-environments.

Taphonomy

Accurate reconstruction of a fossil as an individual and ecologically interactive organism requires a thorough understanding of its preservational history. Paleobiological investigation thus merges with sedimentology, low-temperature geochemistry and, ultimately, the manner in which fossils are retrieved and studied. In the absence of significant biomineralization, Proterozoic fossil taphonomy approximates that of non-mineralizing organisms of Phanerozoic age (Butterfield 1990), but with some important exceptions. Before the Ediacaran–Cambrian radiation of bioturbating metazoans, there would have been a markedly enhanced potential for 'stagnation'-type fossil preservation (Seilacher *et al.* 1985), a feature reflected in the Svanbergfjellet Formation by its preservation of finely laminated sedimentary fabrics (Fig. 3E). Moreover, a predominantly abiological Proterozoic silica cycle in which silica was precipitated as a semi-evaporitic phase in peritidal environments (Maliva *et al.* 1990) provided a unique taphonomic window onto near-shore microbial mats and associated biotas; this view was effectively eliminated by the Cambrian radiation of mineralizing sponges and radiolarians(?) and the transfer of silica deposition to offshore environments. Conversely, a relatively late evolution of recalcitrant, aromatically-based biopolymers (e.g., lignin, sclerotin) would skew the record in favor of Phanerozoic organic-walled fossils.

The preservation of pre-Ediacaran fossils followed a number of taphonomic pathways, including compressions and impressions (molds) in shale, early diagenetic permineralization, and mineral replacement. The Svanbergfjellet Formation is remarkable in that it includes examples of all these taphonomic modes, thus offering an unusually diverse view of late Proterozoic life. A majority of the Svanbergfjellet fossils are composed of original (although now kerogenized) organic carbon and owe their preservation to a coincidence of relative recalcitrance of the original structures and more or less anti-biotic depositional circumstances. These have previously been considered with respect to the organic preservation of non-mineralizing animals (Butterfield 1990) and permineralized land plants and peats (Knoll 1985). The discussion is continued here where it concerns taphonomic processes in the Proterozoic.

Recalcitrance

The preservable constituents of non-mineralizing organisms are clearly in the extracellular biopolymers making up cell walls, sheaths, and cuticles (Hanic & Craigie 1969; Gunnison & Alexander 1975; Butterfield 1990); the negligible preserva-

tion potential of cytoplasmic constituents is reflected in their typical absence or severely atrophied appearance in organically preserved fossils (Figs. 5G, 7A, 7E). During the Proterozoic, the most recalcitrant biopolymers were probably strongly aliphatic, oxygen-poor compounds such as sporopollenin. The widespread occurrence of sporopollenin in modern dinoflagellate cysts and prasinophyte phycomata lends support to the interpretation of most Proterozoic acritarchs as their functional, and possibly taxonomic analogues (Tappan 1980). Sporopollenin, however, is also an important constituent of the *vegetative* cell walls of several chlorophyte algae (e.g., Atkinson *et al.* 1972; Marchant 1977) and points to a more heterogenous pool of potential sources for organic-walled microfossils (Butterfield & Chandler 1992). The marked similarity between Svanbergfjellet *Palaeastrum* n.gen. and extant, sporopollenin-containing *Coelastrum* and *Pediastrum* (Chlorophyta), for example, suggests that this fossil owes its preservation to such 'vegetative' sporopollenin.

Sporopollenin is not the only recalcitrant compound produced by protistan-grade organisms. Biopolymers that satisfy the degradational tests for sporopollenin (resistance to acetolysis and saponification) but with distinct chemical structures are important cell-wall constituents of various chlorophyte algae, including the hydrocarbon-rich *Botryococcus braunii* (Berkaloff *et al.* 1983; Puel *et al.* 1987). These and other 'alghumins' (='algaenans') often appear to be more recalcitrant than even sporopollenin (Philp & Calvin 1976; Hatcher *et al.* 1983; Zelibor *et al.* 1988; Tegelaar *et al.* 1989). While themselves comprising only a minor proportion of the cell wall, such compounds may serve to shield more labile cell wall constituents (e.g., cellulose) by altering or blocking sites of enzyme activity (Alexander 1973; Zelibor *et al.* 1988). Comparably recalcitrant, but chemically distinct biopolymers are also reported from the envelopes (cell walls and sheaths) of some cyanobacteria (e.g., Chalansonnet *et al.* 1987) but, unlike many of their eukaryotic counterparts, these exhibit little structural integrity. The moderate recalcitrance of envelope glycolipids, peptidoglycans, and lipopolysaccharides (Tegelaar *et al.* 1989) must also have contributed to the preservation of Proterozoic cyanobacteria.

It is indeed *relative* rather than extreme recalcitrance that should guide taphonomic analysis, particularly in cases of exceptional preservation where fossil representation is skewed away from a simple view of resistant cysts and spores. Relative recalcitrance is of course dependent upon numerous interacting factors, but will clearly include circumstances of deposition, taxonomy (although in no immediately obvious pattern; Hanic & Craigie 1969; Foree & McCarty 1970; Gunnison & Alexander 1975; Chalansonnet *et al.* 1987; Zelibor *et al.* 1988), and the physiological state of a prospective fossil. For example, very young and very old algal cultures may be particularly resistant to decay (Jewell & McCarty 1971), while some cyanobacteria and algae can remain viable for extended periods if buried while still alive (Gunnison &

Alexander 1975; Birch *et al.* 1983); such a delay in degradation might allow a prospective fossil the grace period necessary for its full isolation in the sediment.

Degradation

All naturally occurring organic compounds are biodegradable (Alexander 1973); preservation of organic-walled fossils thus requires that the normal cycle of decomposition be prematurely and wholly terminated. Prior to the introduction of macrophagous animals such degradation would have been dominated by mechanical fragmentation, photolysis (Kiebar *et al.* 1990), and heterotrophic and autolytic biochemistry.

Geochemical signatures indicate that the biogeochemical cycling of carbon has been active since the early Archean, yet direct evidence of early heterotrophic microbes is rare (Lanier 1988). Svanbergfjellet *Brachypleganon* (Fig. 22J–K) were previously proposed as fossil heterotrophs (Butterfield *et al.* 1988), but a possible autotrophic metabolism cannot be ruled out. The same is true for the *Siphonophycus*-type filaments that commonly line the body cavity of *Valkyria* n.gen. (Fig. 10H). Unambiguous evidence of heterotrophy in the Svanbergfjellet biota occurs as micro-trace fossils in the walls of various acritarchs (Figs. 12H, 13B, 19E). These distinctive circular perforations are clearly not primary, nor could they have been formed through sedimentary compaction since they typically pass through only one wall of the now double-walled compressions. The exact nature of the responsible heterotroph(s) is not known, but the larger perforations (up to 50 μm in diameter in some leiosphaerids) are conceivably the product of colonial, degradative bacteria; the 2–3 μm holes in *Cerebrosphaera buickii* n.sp. (Fig. 12H) appear to have had a distinct origin, possibly from an individual predator or degrader.

While heterotrophic bacteria are central to an operative carbon cycle, a number of studies suggest that particulate organic carbon (POC), the 'stuff' of organic-walled fossils, serves as a particularly poor microbial substrate (Burns 1979; Fallon & Brock 1979; Karl *et al.* 1988). Microbial biodegradation appears instead to be centered on the very finest POC and/or dissolved organic carbon (Fallon & Brock 1979). Indeed, axenic (bacteria-free) cultures of degrading algae show much the same degree of particle reduction as those seeded with heterotrophic bacteria (Foree & McCarty 1970). The clear implication is that organisms are inherently self-destructing, presumably as a result of autolytic enzymes. Autolytic degradation results from the activation of enzymes involved in the remodelling or removal of cell walls as they accommodate growth, division, and the release of gametes and zygotes (Ferris *et al.* 1988; Matsuda 1988; Butterfield 1990). The specific capacity of autolysins to degrade otherwise recalcitrant cell walls, their immediate post-mortem availability, and their potential accumulation in sediments (Burns 1979; Matsuda 1988) suggests they represent a funda-

mental barrier to organic-walled fossil preservation. Indeed, the importance of autolytic enzymes to initial biodegradation may help to explain the preferential preservation of cyanobacterial sheaths over their seemingly more recalcitrant cell walls (Golubic & Barghoorn 1977; Bauld 1981) – where the latter are subject to autolytic restructuring, mucilaginous sheaths are simply vacated or plastically re-formed and are therefore not accompanied by specific autolysins.

Preservation

Early isolation by burial is necessary, though not in itself sufficient, for the preservation of organic-walled fossils. The exceptional preservation often encountered in silicified 'flake conglomerates' (e.g., Knoll 1982; Mendelson & Schopf 1982; present study) suggests a bias in favor of rapid sedimentation, particularly when removed from sites of concentrated biological activity. By the same token, planktic microorganisms might be systematically under-represented in the fossil record due to negligible sinking rates. Accelerated deposition of such forms is nevertheless effected through attachment to larger sinking particles (Lochte & Turley 1988) including, in siliciclastic environments, the mutual flocculation of plankton and clays (Avnimelech *et al.* 1982). In this latter case, particle formation is dependent upon available cations (divalent cations more readily induce flocculation than monovalent cations), total organic carbon (high TOC inhibits flocculation; Theng 1979), and the physiological state of interacting organisms (e.g., Dawson *et al.* 1981; Fattom & Shilo 1984).

Once deposited, sediment interment fundamentally alters the degradational context of a prospective fossil. Fine-grained (or very early mineralized) sediments can, for example, prevent the access of heterotrophic microbes (Fontes *et al.* 1991) and/or their respective electron acceptors (e.g., oxygen; Revsbech *et al.* 1980), while the sealed-in metabolites of early biodegradation may inhibit further heterotrophic activity (Butterfield 1990). However, as discussed above, exclusion of microbial activity is not synonymous with the cessation of biodegradation; nor is degradation significantly impeded by the imposition of anaerobic conditions (Foree & McCarty 1970; Fallon & Brock 1979; Allison 1988; Butterfield 1990) or moderate reductions in temperature (Foree & McCarty 1970; Kidwell & Baumiller 1990; Wiebe *et al.* 1992). Preservation of organic-walled fossils requires a rapid and profound interference with all enzyme-mediated biochemistry.

Organic-walled fossils in the Svanbergfjellet Formation are occasionally preserved in dolomite microspar (Fig. 23K) or permineralized in large euhedral dolomite crystals (Fig. 3G); most, however, occur either as compressions in shale or as permineralizations in early diagenetic silica. These principal taphonomic modes differ fundamentally in process and overall fossil constitution. Silicified fossil example, seem not to be extractable by acid dissolution chert matrix and owe their integrity more to the marked stability of chert than to the actual preservation of organic material. By contrast, the generally coherent and acid-extractable fossils in shale derive from the exceptional preservation of organic structure *per se.*

The marked anti-enzymatic activity of particular clay-organic systems has long been recognized by soil scientists (Burns 1979; Theng 1979) and appears to be an important mechanism in the organic preservation of shale-hosted fossils (Butterfield 1990). Degradative enzymes are potentially inhibited by adsorption onto and within clay minerals, with the overall efficacy dependent upon clay type, available exchange cations, pH, and total organic carbon (TOC); maximum preservation potential is expected to occur with expanding, montmorillonite-type clays, monovalent exchange cations, acid pH, and relatively low TOC (Butterfield 1990). Other combinations will also undoubtedly induce fossilization; however, both the fossil record and theoretical considerations point to the central importance of low organic-carbon-to-sediment ratios (low TOC). In the Svanbergfjellet shales, loss of fossil definition and increased inter-fossil fusion is related directly to increasing TOC, such that organic-rich horizons ($\geq 1.5\%$ TOC) preserve only extremely robust taxa such as *Chuaria* and *Tawuia*.

It is with respect to TOC that fossil preservation in early diagenetic silica departs most notably from that in shales. Such permineralization appears to result from a particular affinity of dissolved monosilicic acids with the exposed functional groups of degrading organic compounds, hence its particular association with *high* TOC (Leo & Barghoorn 1976; Knoll 1985); indeed, it is the *enhanced* chemical reactivity of the system, rather than its inhibition (as with shales), that leads to increased preservation potential in chert. Together with a predominantly peritidal locus of early diagenetic silica during the Proterozoic (Maliva *et al.* 1990), this basic disparity of process strongly influenced the environments and ecosystems sampled by the two taphonomic modes. In broadest terms, early silica permineralization is inversely related to sediment supply, hence its prevalence in supratidal microbialites where mat material might constitute close to 100% of the original volume, and its virtual absence from columnar (i.e. sediment-rich) stromatolites. Conversely, the positive correlation between exceptional preservation and sediment input in shale facies tends to emphasize less restricted paleoenvironments and consequently more diverse paleoecosystems. These two taphonomic systems are of course not entirely exclusive, and they broadly intersect in shallow subtidal environments (p. 8).

Mineralization. – Fossil preservation through early-diagenetic mineral replacement is an important taphonomic mode for Phanerozoic fossil *Lagerstätten* (Allison 1988), yet it represents only a minor component of the Proterozoic record (e.g., Wang *et al.* 1983; Knoll *et al.* 1991). In the Svanbergfjellet Formation it accounts for two filamentous but otherwise problematic forms. In one instance, bedding-parallel thin sections of an Algal Dolomite Member shale (sample 86-G-30) reveal abundant, often thickly entangled cylindrical structures that petrographic and EDAX analysis show to be composed of a yellow, anisotropic, iron-rich mineral, possibly goethite (Fig. 26C–F). The occasional inclusion of intact organic-walled filaments attests to the biogenicity of these fossils; however, the mineral overgrowths have significantly distorted the original morphology (see systematic section). The other occurrence is of abundant, *vertically*-oriented and branched(?) filamentous structures in shale-hosted, early-diagenetic apatite nodules of the Lower Dolomite Member (sample SV-3; Fig. 23F–G). The preserved lamination in the nodules appears to be that of the uncompressed shale and thus provides a unique view of the third (i.e. vertical) dimension in fine-grained siliciclastic facies. While the taxonomic affiliation of these structures is unclear, their formation is difficult to explain as other than biological.

Evidence of primary biomineralization among the Svanbergfjellet fossils is equivocal. *Siphonophycus*-like filaments in the interstitial spaces of a silicified microbialite grainstone (86-G-15) are marked by intervals of densely packed, sub-micrometer-sized particles that create a distinctive banded pattern (Fig. 23B–D) broadly reminiscent of the biologically-induced biomineralization of some extant filamentous microbes (e.g., *Scytonema julianum*). These particles, however, do not now show any optical activity beyond that of the replacive chert, and they conceivably represent a simple organic encrustation. Similarly banded filaments occur in Chichkan Suite, Kazakhstan (Sovetov & Schenfil 1977, Fig. 2e).

Analysis of fossils

Taphonomy can be addressed deductively as a question of materials and process, such as emphasized in the above discussion; alternatively, the analysis can begin with a fossil and work back to reconstruct the living organism. Under this latter approach, interpretation of fossils will depend largely upon taphonomic studies on modern analogues (i.e. actualistic experimentation) or a comparison of fossil taxa under differing preservational modes (e.g., contemporaneous shale- and chert-hosted fossils). Moreover, secondary taphonomic features may themselves provide significant information about primary structure, such as its relative plasticity or general constitution (Butterfield 1990). The two approaches are of course not mutually exclusive, and the mental exercise of laying on a taphonomic overprint will surely intersect with attempts to peel it back from fossil material.

Thickness vs. opacity. – Wall thickness is commonly referred to in taxonomic discussions of organic-walled microfossils, although it is sure to be under some degree of taphonomic control. Thus, the thick extracellular sheaths that typify many permineralized fossils are invariably collapsed as compressions, and, in some cases, appear to be reduced to a simple shagrinate surface texture (e.g., *Germinosphaera jankauskasii* n.sp., *Trachyhystrichosphaera aimika*, *Digitus adumbratus* n.sp.). The thickness of true cell wall appears to be less affected by taphonomic mode; however, it too will certainly vary with the overall quality of preservation. Despite this taphonomic factor, the original nature of fossil cell walls can sometimes be inferred independently through an examination of enclosing siliciclastic sediments. Thus, the thick and resilient walls of *Chuaria*, *Tawuia* and *Cerebrosphaera* n.gen. may be recognized by their prominent (and potentially diagnostic) impressions onto otherwise flat bedding planes, even in the absence of original organic material (e.g., Fig. 8E, G, H; Gussow 1975). By contrast, originally thin-walled or otherwise insubstantial structures leave little or no sedimentary imprint (Fig. 8A, F). The thickness of intact wall material can of course be measured directly under SEM (Figs. 13H, 23H).

Thickness might also be inferred from the relative opacity of organic structures; however, a systematic carbonization of organic-walled fossils with increasing metamorphic grade clearly rules out the universal application of such a measure. There are also clear taxonomic and histological differences in fossil opacity (Butterfield 1990). In the Svanbergfjellet, for example, *Chuaria*, *Tawuia*, and *Cerebrosphaera* n.gen. are the only fossils regularly opaque to the light source of a standard laboratory microscope. Granted, these taxa are all relatively thick-walled, but close examination reveals that this is not the direct cause of their optical density. A slide-mounted, ca. 3 μm thick fragment of *C. circularis* failed to become translucent through incremental polishing until it was all but removed; the material making up the wall is inherently dark. Similarly, the outer(?) ca. 1 μm thick layer of *Tawuia* is opaque, yet a lightly bonded 3.5 μm thick inner(?) layer remains remarkably translucent (Figs. 8D, 23H). Pronounced histological differences in opacity are also noted in the various structures of *Valkyria* n.gen., *Proterocladus* n.gen., *Palaeastrum* n.gen., and *Trachyhystrichosphaera polaris* n.sp. Thus, while relative opacity does not necessarily translate to relative thickness, it may be of some histological and, thereby, taxonomic significance.

Compression. – The projection of three dimensional structure onto the two dimensions of a bedding plane distinguishes shale-hosted carbonaceous fossils from most other preservational modes. Under compression, organisms with an originally circular cross-section might be expected to increase their overall dimensions by a factor of $\pi/2$ (=57%) (Hofmann & Aitken 1979; Zhang Z. 1982; Schopf 1992); such, however, seems rarely to be the case. Comparison of two- and three-dimensionally preserved graptolites (Briggs & Williams 1982) and actualistic compression studies on various three-dimensional 'fossils' (Harris 1974; Rex & Chaloner 1983; Rex 1983) reveals that most flattened fossils can be viewed as relatively undistorted two-dimensional images of the original structures, the equivalent of photographs. Per-

mineralized and compression fossils will thus be broadly equivalent, at least with respect to absolute dimensions.

The most obvious way to accommodate the surplus wall of the third dimension during flatting is through folding (Harris 1974, type 3 failure), as is apparent in many, perhaps most, acritarchs. Alternatively, the excess may be taken up through a plastic thickening of the walls (type 4 failure), a mode suggested here for the probably quite plastic sheaths of cyanobacteria. Unfolded compression fossils might also deform plastically through lateral expansion (type 5 failure); however, the difficulty in achieving this behavior experimentally (Harris 1974) and the closely comparable diameters of *Siphonophycus robustum* and *S. typicum* n.comb. in Svanbergfjellet cherts and shales militate against such an interpretation. In any event, collapse due to decay (which does not involve lateral expansion) will significantly precede sedimentary compaction in most depositional environments (Conway Morris 1979; Briggs & Williams 1981).

Some Svanbergfjellet fossils do show evidence of compression-induced distortion. The robust, non-folding walls of *Cerebrosphaera* n.gen. typically accommodated flattening through the formation of radial fractures (Fig. 12C–F), a habit betraying the primary rigidity of its walls, and accompanied by substantial lateral expansion (Harris 1974, type 2 failure). In a comparable manner, the single radial split that characterizes many *Leiosphaeridia wimanii* n.comb. acritarchs results in an artificially elongated (ellipsoidal) compression fossil (Fig. 13E–F). Shape distortion may also occur in unsplit folded compressions, but the mean diameter is expected to approximate that of the original structure.

The points raised here of course deal with a minority of the processes that influence the occurrence and final appearance of a fossil. Further study, especially of fossil *Lagerstätten* and their particular depositional, biogeochemical, and temporal contexts, should significantly enhance paleobiological resolution. In summary, a detailed understanding of taphonomic processes is necessary to appreciate the form and degree of post-mortem information loss, to assess accurately that which remains, and to serve as a guide in the search for further fossil occurrences.

Taxonomy

Systematic taxonomy is the framework supporting most biological and paleobiological investigation. Among living organisms it is based largely on degree of phenotypic similarity with a 'type' (and a range of intraspecific variation), despite the widespread acceptance of the biological species concept. The same is true (necessarily) for fossil species, but here a 'natural' classification may become obfuscated by taphonomic alteration and a dearth of close modern analogues, especially in the case of Proterozoic forms. Insofar as we are interested in the fossil record as a guide to evolutionary history and paleobiology in general, every attempt should be made to look upon fossils as real organisms preserved in various stages of ontogeny and postmortem decay.

Significant details of Proterozoic fossil biology can often be determined through a consideration of functional morphology and preserved 'behavior' (e.g., the coloniality of *Brachypleganon* or the phototropism of *Polybessurus* and *Siphonophycus*), as well as the intraspecific variation revealed by large populations. Thus, on the basis of their unique processes and a clear morphological continuum, nine previously named 'species' of the acanthomorphic acritarch *Trachyhystrichosphaera* are now recognized as developmental and/or ecophenotypic variants of a single biological entity, even though the higher-order taxonomy remains unclear. Conversely, a judicious recourse to type material may help to resolve relatively meaningless 'wastebasket' taxa into well-defined groups (e.g., *Chuaria*/*Leiosphaeridia*/*Cerebrosphaera* n.gen.). In a few instances, fossils share a sufficient complement of characters with living analogues (homologues) that they can be treated under a fully natural and modern taxonomy, as is the case for the cyanobacteria *Polybessurus*, *Eoentophysalis* and *Obruchevella*, and the Svanbergfjellet siphonocladaleans in *Proterocladus* n.gen.

An artificial 'form taxonomy' may be unavoidable for very simple Proterozoic fossils such as spheroids and filaments. Such treatment will clearly conflate disparate natural taxa (compare, for example, modern *Oscillatoria* with *Beggiatoa*, or *Phormidium* with *Chloroflexus*); however, broadly meaningful groups can often be defined (e.g., regularly recurring size-frequency distributions among *Leiosphaeridia* or *Siphonophycus* – see systematic section). On the other hand, a taxonomic recognition of *taphonomic* variants needlessly over-estimates paleodiversity; the practice of treating silicified and shale-hosted fossils under separate taxonomies, for example, is now recognized as largely misleading (Knoll 1984; Pjatiletov 1988; Jankauskas *et al.* 1989; Knoll *et al.* 1991). To be useful, fossil taxa must accommodate a degree of natural variation (e.g., Shukovsky & Halfen 1976; Haxo *et al.* 1987) and taphonomic alteration (e.g., Golubic & Barghoorn 1977; Horodyski *et al.* 1977) comparable to that observed in their modern analogues. A relatively conservative taxonomic policy has been taken in the present study under the assumption that it will at least provide a reliable estimate of *minimum* diversity.

Filamentous microfossils of prokaryotic aspect comprise one of a number of taxonomic quandaries in Proterozoic taxonomy; several hundred 'species' have been erected to describe a relatively limited range of morphology. Their abundance and diverse preservation in the Svanbergfjellet Formation offer an excellent opportunity to review their overall taxonomy. As strict form taxa, eight basic forms (genera) can be recognized among the Svanbergfjellet filaments: (1) *Oscillatoriopsis* – unsheathed cellular trichomes with cell width ≥ cell length; (2) *Cyanonema* – unsheathed cellular trichomes with cell width < cell length; (3) *Veteronostocale* – unsheathed cellular trichomes severely constricted at the intercellular septa; (4) *Palaeolyngbya* – cellular trichomes within a single outer sheath; (5) *Tortunema* – pseudoseptate filaments; (6) *Cephalonyx* – 'pseudocellular' filaments; (7) *Rugosoopsis* – cellular or acellular filaments with a double outer sheath, the outer one having a transverse fabric; and (8) *Siphonophycus* – smooth-walled filamentous sheaths. Major synonymies and systematic revisions of these taxa are presented in the 'Systematic paleontology' section.

Multicellularity

The advent of eukaryotic multicellularity represents one of the fundamental advances in the history of life on Earth. How and why some organisms came to be constructed of numerous integrated units is an ongoing and fascinating subject of investigation (Sharp 1934; Cavalier-Smith 1978; Buss 1987; Kauffman 1987; Bonner 1988; Kaplan & Hagemann 1991). A significantly pre-Cambrian record of multicellular life is suggested by both molecular phylogenetic studies (Runnegar 1982; Sogin 1989) and basic evolutionary theory; however, the detailed paleontological evidence necessary to confirm and constrain this conjecture has remained largely elusive. The Svanbergfjellet shales (especially those of the Algal Dolomite Member) offer a new and substantially enhanced view of multicellular life 150–200 Ma before the Ediacaran–Cambrian radiations of large animals.

In addition to fossils, a search for pre-Ediacaran multicellularity requires a strict definition of the term and an appreciation of how such a habit might be recognized based solely on fossil morphology. In the first instance, 'multicellular', as here applied, refers only to eukaryotic organisms; prokaryotes can indeed be constructed of multiple cells, but such interaction has led to only limited morphological differentiation or evolutionary innovation (Awramik & Valentine 1985). Fundamentally unicellular organisms (prokaryotic or eukaryotic) that have simply agglomerated, or divided but failed to separate, are likewise excluded from the discussion (e.g., *Ostiana*). Conversely, demonstrably coenocytic (i.e. multinucleate but non-cellular) forms are included here under the assumption that they represent either a primitive or derived state of multicellularity. Relatively large size may serve as an accessory criterion for multicellularity but is neither necessary or sufficient for identifying the condition. True multicellular fossils are expected to exhibit signs of cellular differentiation, intercellular communication, higher-order (i.e. emergent) structure, and/or morphological comparison with extant organisms that are accepted as being multicellular.

Simple multicellularity

Of the eight taxa of multicellular organisms preserved in the Svanbergfjellet Formation, the simplest is *Palaeastrum* n.gen. (Fig. 5A–C). At first glance simply an agglomeration of spheroidal unicells, its distinct and regular differentiation of intercellular attachment structures (plaques) shows *Palaeastrum* to have had a colonial–coenobial grade of multicellular organization closely comparable to the coenobia of some extant chlorococcalean green algae (e.g., *Pediastrum* or *Coelastrum*; Fig. 5D). By contrast, species of *Germinosphaera* are not multiple-celled and, strictly speaking, are classified among the Acritarcha. The random arrangement of their one to several open-ended and occasionally branched tubular processes nevertheless points to an active growth habit, similar to that of germinating zoospores in some modern filamentous protists. The xanthophyte alga *Vaucheria*, for example, reproduces asexually by means of large (ca. 100 µm) spheroidal zoospores (actually multinucleate coenocytes) that germinate one or more filamentous, sometimes branched primordia (Fig. 16G) in a fashion indistinguishable from that seen in *G. fibrilla* n.comb. (Fig. 17). The known (cf. Oomycota) and probable convergence upon this relatively simple habit precludes a positive assignment of germinosphaerids to any particular protistan lineage; their principal significance lies in documenting a fully coenocytic grade of multicellularity in the Svanbergfjellet assemblage.

Filamentous Chlorophyta (green algae). – Some Svanbergfjellet fossils stand as true multicellular protists without equivocation. The uniseriate filaments of *Proterocladus* n.gen. are constructed of multiple cells separated by differentiated intercellular septa, express a higher order morphological complexity through lateral branching, and can be quite large (a specimen of *P. major* n.sp. extends for ca. 1 cm in bedding-parallel thin-section). Uniseriate cellular filaments are of course distributed widely among protistan-grade organisms; however, these fossils are distinguished by unusually long and markedly irregular cell lengths. In combination with various details of branching, septum formation, and apparent reproductive structures, such a pattern is particularly characteristic, indeed diagnostic, of the modern siphonocladalean chlorophyte *Cladophoropsis*. Species of *Proterocladus* are consequently assigned to the chlorophyte order Siphonocladales (cellular, but multinucleate Ulvophyceae).

Cladophoropsis owes its distinctive cellular habit to a unique 'segregative cell division' wherein the cytoplasm of an essentially coenocytic (i.e. multinucleate) filament cleaves at various intervals, lays down an intervening septum, and initiates a single lateral branch at the uppermost end of the newly defined cell (Børgesen 1913). Septum formation may occur without branching, and branches may themselves undergo further segregative cell division; however, a branch typically remains in full cytoplasmic communication with its parent cell, i.e. a septum is not formed below the branching point in the parental axis. This is very much the pattern observed in *Proterocladus hermannae* n.sp. By contrast, both *P. major* n.sp. and *P. minor* n.sp. occasionally exhibit sub-branch septum formation (Figs. 6F, 7A), and the former includes rare specimens with multiple lateral branches on a

single cell (Fig. 6B, J) or with branches not immediately associated with a septum. In these respects *P. major* and *P. minor* would appear to be more closely affiliated with the diverse branching habits of *Cladophora* (cf. Hoek 1984) than with *Cladophoropsis* itself.

A distinction has been made between those chlorophytes that undergo segregative cell division (Siphonocladales) and the more regularly dividing (and divided) forms, the Cladophorales. Ultrastructural and molecular analyses now show these features to be non-diagnostic at the ordinal level (O'Kelly & Floyd 1984; Hoek 1984; Zechman *et al.* 1990) and the Cladophorales is thus subsumed into the Siphonocladales (O'Kelly & Floyd 1984, p. 135). As a consequence, the mosaic of *Cladophoropsis* and *Cladophora* characters exhibited by *Proterocladus* n.gen. become fully accommodated by the order. Indeed, Hoek (1984) argues on morphological grounds that modern *Cladophoropsis* and *Cladophora* are sufficiently alike to be merged as a single genus. In any event, the large cell volumes and clear siphonocladalean affiliations of *Proterocladus* can be used to infer the multinucleate, semi-coenocytic nature of its cells.

Also deriving from the molecular data is evidence that the Class Ulvophyceae Mattox & Stewart 1984 is polyphyletic (Zechman *et al.* 1990). Interestingly, the monophyletic subset of the Ulvophyceae that includes the Siphonocladales is very much that of the Class 'Bryopsidophyceae' (Round 1963), which had previously been delineated on largely morphological criteria. The possibility that non-cytoplasmic constituents can accurately reflect higher order taxonomy is especially reassuring to paleontologists.

Complex multicellularity

The multicellular fossils discussed thus far are notable for their lack of significant cellular differentiation, surely the necessary prerequisite to the evolution of tissue-grade plants, animals, and, in some sense, fungi. *Complex* multicellular fossils are a conspicuous, if taxonomically problematic, component of the Svanbergfjellet assemblage and provide a unique view onto the status of early intraorganismal specialization. The most common (and complex) such form is *Valkyria* n.gen., a large (up to 1 mm long) sausage-shaped fossil with peculiarly lobate lateral 'axes', a medial stripe, and a variety of associated vesicular structures (Figs. 9, 10, 11); at least six discrete 'cell types' are recognizably preserved (see systematic section). Using Bonner's (1988, p. 128) 'number of cell types' as a measure of relative complexity or 'grade of organization', *Valkyria* rates as *at least* as complex as the most differentiated modern algae or fungi. Moreover, it is not immediately clear whether some or all of these structures are differentiated single cells, or the product of a number of specialized cells. In the latter case they would then be of a tissue grade of organization, and *Valkyria* would represent a level of complexity otherwise not recognized until the appearance of Ediacaran faunas, some 150 Ma later.

The problem of cell versus tissue recognition is also encountered in the problematic macrofossil *Pseudotawuia* n.gen. with its terminal pair of large reniform structures (Fig. 8A). It is possible that the various components of *Pseudotawuia* represent single, albeit highly differentiated, cells; however, the combination of both large size and apparent bilateral symmetry point intriguingly (although not conclusively) in the direction of tissue-grade complexity and metazoan affiliation. In this regard it is important to recall that most animal cells and tissues are not preservable as organic-walled fossils and that preservation will be limited largely to cuticle and other secreted, extracellular 'tissues' (Butterfield 1990).

The macroscopic problematicum *Tawuia* occurs worldwide in late Proterozoic sediments (Hofmann 1985a, b), including those of the Svanbergfjellet Formation. It does not exhibit the localized differentiation of *Pseudotawuia* n.gen. (the vague terminal structures present in some *Tawuia* are mechanically induced; Hofmann 1985a). However, close examination of well preserved Svanbergfjellet material shows it to have a distinctive bi-layered wall construction: a ca. 1 μm thick, opaque, brittle layer is lightly bonded to a ca. 3.5 μm thick, translucent and flexible inner(?) stratum (Figs. 8D, 23H). Along with its considerable size, such marked histological differentiation establishes *Tawuia* as a complex and probably multicellular organism. Absence of any preserved cellularity suggests that *Tawuia* cells were unwalled, or that the organism was fundamentally coenocytic; the very robust, and apparently continuous, wall argues against a metazoan affiliation.

Other Proterozoic occurrences

Multicellular fossils of varying degrees of taxonomic and temporal resolution are preserved elsewhere in the Proterozoic. Uniseriate and multiseriate filaments from the ca. 725–1250 Ma Hunting Formation in arctic Canada have been identified as bangiophyte red algae (Butterfield *et al.* 1990), primarily on the recognition of diagnostic cell division patterns. The branched, essentially nonseptate filaments of *Palaeovaucheria clavata* Hermann, 1981(a), from the Late Riphean Lakhanda suite of Siberia were clearly multinucleate eukaryotes and are tentatively accepted as a *Vaucheria*-like alga (Chromophyta), despite the indistinguishable morphology of some Oomycota. *Palaeosiphonella* Licari, 1978, is superficially comparable, but as it occurs in mat-like associations in stromatolitic facies it is perhaps better interpreted as the false-branched sheath of a cyanobacterium; this is clearly the case for the much smaller branched filaments of *Ramivaginalis* Nyberg & Schopf, 1984.

Macroscopic and apparently coenocytic unbranched filaments are reported from shales ca. 1100 Ma old in India (*Katnia* Tandon & Kumar, 1977), ca. 1400 Ma in North China and Montana (*Grypania* Walter *et al.*, 1976; Du *et al.* 1986; Walter *et al.* 1990), and ca. 1800 Ma in North China

(Hofmann & Chen 1981), as well as from ca. 2100 Ma BIF in Michigan (Han & Runnegar 1992). Large, septate filaments with very large constituent cells are also reported from 1700–1800 Ma shales of North China (*Qingshania* Yan, 1989). While certainly eukaryotic, the taxonomic affiliation of these fossils remains speculative. The same is true for the widely distributed tawuiids and longfengshaniids of the pre-Vendian Neoproterozoic (Hofmann 1985a, b) and a variety of putative algal (Timofeev *et al.* 1976; Timofeev & Hermann 1979; Jankauskas *et al.* 1989) and fungal (Hermann 1979) remains in the Lakhanda sequence. Problematic bedding-plane markings in middle Proterozoic sandstones of Western Australia (Grey & Williams 1990) and Montana (Horodyski 1982) are conceivably the imprints of a relatively large seaweed, and a possible mineralized metaphyte is reported from late Proterozoic carbonates in southeastern California (Horodyski & Mankiewicz 1990). Fossils and ichnofossils of multicellular organisms of course abound in latest Proterozoic (Ediacaran) rocks, although probable algal forms are relatively few and morphologically simple (e.g., Zhu & Chen 1984; Gnilovskaya 1988; Zhang 1989; Grant *et al.* 1991; Chen & Xiao 1991; Zhang & Yuan 1992).

Evolutionary implications

While admittedly patchy, the pre-Ediacaran record of multicellular fossils has some interesting aspects. For example, those forms that can be reasonably classified into extant taxonomic groups are all algal. And, with the addition of the Svanbergfjellet assemblage, there is now strong paleontological evidence for the presence of all three principal algal clades, the Rhodophyta (Butterfield *et al.* 1990), Chromophyta (Hermann 1981b), and Chlorophyta (present study), by at least 750 Ma ago. Moreover, each of these three lineages had independently evolved a true multicellular organization by that time, if only of a relatively simple grade. Conspicuously absent is evidence of parenchymatous or pseudo-parenchymatous construction with its inherent capacity for producing large and elaborate algal structures such as appear in the early Paleozoic (e.g., Walcott 1919). Indeed, even the simple, regularly septate filaments that dominate (at least numerically) most modern algal communities are rare in the Proterozoic and entirely absent in the Svanbergfjellet assemblage. Although data are limited, the most common 'grade' of pre-Ediacaran multicellularity appears to have been either coenocytic (e.g., *Palaeovaucheria, Germinosphaera, Tawuia, Grypania* Walter *et al.*, 1976; *Longfengshania* Du, 1982, *Majaphyton* Timofeev & Hermann, 1976, *Archaeoclada* Hermann, 1989, *Variaclada* Hermann, 1989, *Caudosphaera* Hermann & Timofeev, 1989) or semi-coenocytic (i.e. very large, multinucleate cells such as occur in *Proterocladus* n.gen., *Qingshania* Yan, 1989, and *Eosolena* Hermann, 1985). The exceptions to this generalization include the agglomerative multicellularity of *Palaeastrum* n.sp. and *Eosaccharomyces* Hermann, 1979, the intercalary cell division of the Canadian bangiophyte, (Butterfield *et al.* 1990), and the possibly prokaryotic '*Gunflintia*' *barghoornii* Maithy, 1975, *Trachythrichoides* Hermann, 1976, and *Lomentunella* Hermann, 1981. Notably, all probable metaphytes of middle and early Proterozoic age appear to have been coenocytic (e.g., Walter *et al.* 1976; Tandon & Kumar 1977; Hofmann & Chen 1981; Du *et al.* 1986; Walter *et al.* 1990; Han & Runnegar 1992) or, rarely, semi-coenocytic (e.g., Yan 1989).

The possibility of an evolutionary sequence leading from coenocytic to multiple-celled multicellular organisms is particularly intriguing in light of recent (Kaplan & Hagemann 1991) and not so recent (Sharp 1934) inquiries into the applicability of conventional 'cell theory' to the understanding of multicellularity. In essence, cell theory holds that a multicellular body is (both developmentally and evolutionarily) an aggregate of unicells, each of the same rank and the equivalent of a unicellular organism. The alternative 'organismal theory' would argue that it is the whole multicellular body that equates with a unicellular organism, and that large size and morphological differentiation are independent of, and are likely to have preceded cellularity. Such independence is supported by experimental studies wherein growth and development may proceed even when cell division has been suppressed (Sharp 1934; Kaplan & Hagemann 1991), and by the natural occurrence of diverse coenocytic organisms that converge upon complex, cellular-grade and tissue-grade forms (e.g., the coenocytic chlorophytes *Bryopsis* and *Caulerpa*). Early coenocytes would have generally come under various selective pressures for internal partitioning (e.g., mechanical support; Kaplan & Hagemann 1991), hence the typical situation of large organisms being constructed of multiple cells. Because a primitive cell-division program is unlikely to have closely synchronized cytokinesis (septation) and karyogenesis (mitosis), the cells of early multiple-celled multicellular organisms are expected, under this evolutionary scenario, to have been multinucleate and of irregular size (such as found in the Siphonocladales). *Regular* cellularity would then arise as developmental programs became more refined. As discussed above, the limited fossil record suggests that such refinements may have appeared relatively late in the Proterozoic.

A coenocytic to cellular evolutionary sequence is supported by molecular phylogenetic analysis (rRNA) within at least one major branch of the green algae. The consensus phylogenetic tree of Zechman *et al.* (1990, Fig. 3) shows that regularly septate *Cladophora* and *Chaetomorpha* form a sister group to irregularly septate *Cladophoropsis* and that the Siphonocladales as a whole is sister group to the coenocytic Caulerpales and Dasycladales (with a probable coenocytic common ancestor). Whether this pattern holds true for other multicellular lineages awaits further analysis in groups that retain both coenocytic and cellular forms. Among the fungi, for example, it is the regularly septate and strictly dikaryotic Basidiomycota that are thought to be the most highly derived, followed by the often irregularly multinucle-

ate Ascomycota; coenocytic Zygomycota and Chytridiomycota appear to represent the most primitive state (Tehler 1988; Bruns *et al.* 1991; Bowman *et al.* 1992). In any event, the early fossil record discounts the often tacit assumption that coenocytic forms are necessarily derived from multiple-celled, multicellular ancestors (e.g., Sears & Brawley 1982).

There are, of course, other means of becoming multicellular. Inherently unicellular organisms can agglomerate and fuse to form a higher-order structure (e.g., the cellular slime molds), or fail to separate following cell division to yield structured multicellular colonies (e.g., the coenobial green algae, including *Volvox* and the Svanbergfjellet fossil *Palaeastrum* n.gen.; Fig. 5A–C). The question is whether such interactions ever gave rise to subsequent evolutionary innovations, as cell theory would hold, or whether they represent an independent and essentially dead-end strategy for increasing size without a concomitant increase in complexity (Bonner 1988). Particularly in the light of accruing fossil evidence, 'it is not at all clear that these afford the key to the evolution of [multicellular] organisms in general' (Sharp 1934, p. 22). Agglomeration/non-separation also appears to be the principal mode of 'multicellularity' among prokaryotes (e.g., myxobacteria, cyanobacteria, actinomycetes) and may account for their relatively limited morphological evolution (cf. Awramik & Valentine 1985). Indeed, it may be the acquisition of an organism-level gene complex (e.g., the organizational homeobox genes), with or without accompanying multi-cellularity, that most accurately defines what we mean by 'true' multicellularity (cf. Slack *et al.* 1993).

Metazoans(?)

The early and broadly contemporaneous fossil occurrence of rhodophyte, chromophyte, and chlorophyte algae is in keeping with various molecular phylogenetic analyses suggesting that these three clades diverged from a common ancestor at or around the same time (Perasso *et al.* 1989; Bhattacharya *et al.* 1990). Interestingly, the molecular data at this level of resolution do not readily distinguish the divergence of the metazoan clade(s) (Sogin *et al.* 1989; Hendriks *et al.* 1991); animals too are likely to have had a significant pre-Ediacaran history. With respect to the fossil record, putative pre-Ediacaran metazoans have been limited largely to various vermiform compressions in the late Proterozoic Liulaobei and Jiuliqiao Formations of China (e.g., *Sinosabellidites* Zheng, 1980; *Pararenicola* Wang, 1982; *Protoarenicola* Wang, 1982; *Huainella* Wang, 1982; *Paleolina tortuosa* Wang, 1982), but absence of any diagnostic metazoan features precludes a final assessment of these fossils. The same is true for the Svanbergfjellet fossils *Valkyria* n.gen. and *Pseudotawuia* n.gen., although these do offer the additional feature of clearly differentiated cells or tissues (Figs. 8A, 9, 10, 11).

Because non-mineralized metazoan body fossils are rarely preserved and are subject to ambiguous interpretation, the search for early animal life has come to be directed at the potential trace-fossil record, a program that may be flawed on at least two counts. First, the formation of sediment-displacing trace fossils would imply the existence of a relatively sophisticated locomotory apparatus such as a hydraulic skeleton, yet the earliest animals were likely to have been acoelomate and relatively lethargic. Secondly, if early *energetic* animals were small meiofauna, they will have left little if any perceptible sedimentary traces (cf. Farmer 1992). The undisrupted lamination of most pre-Ediacaran shales (e.g., Fig. 3E) thus does not establish an absence of metazoans, and the more promising approach for documenting early animal history probably lies in the discovery and careful interpretation of body fossils. Even here, however, it will be limited to the first appearance of cuticle and/or other relatively recalcitrant constituents.

Acritarchs

A significant proportion of the Svanbergfjellet fossils are solitary organic-walled vesicles of indeterminate taxonomic affinity – acritarchs. Most are likely to be the cysts, spores, or vegetative unicells of various eukaryotic algae, although prokaryotes, protozoans, or even metazoans (e.g., Kuc 1972) might also be represented. The most common acritarchs in the Svanbergfjellet Formation, as for the Proterozoic in general, are simple, relatively thin-walled spheroids (leiosphaerids) ranging from a few micrometers to over a millimeter in diameter; their classification is based on broad modalities in size-frequency distribution (Fig. 4; Jankauskas *et al.* 1989). While of some practical value, such taxonomy is inherently artificial, as originally indistinguishable vesicles would have developed into such diverse forms as *Germinosphaera*, *Osculosphaera* n.gen., *Pterospermopsimorpha*, and *Trachyhystrichosphaera*. Additional characters may be resolved if such forms are examined *in situ* (e.g., in petrographic thin section), such as recurrent associations or bedding-plane distribution. For example, where some leiosphaerids have a random distribution on shale bedding planes, implying a planktic origin, others occur as discrete localized populations suggestive of benthic growth (Butterfield & Chandler 1992).

A conspicuous fraction of Svanbergfjellet acritarchs feature processes or other distinctive surface ornamentation and are therefore amenable to a considerably more precise, albeit 'form', taxonomy. Ten such morphologically complex taxa have been recorded in shales and cherts of the formation. As in other Neoproterozoic successions, they typically occur in shallow, open-water environments (Butterfield & Chandler 1992) with occasional transport into more restricted (e.g., Fig. 15) nearshore facies.

Biostratigraphy

The fossil record serves its greatest practical role by providing information on the relative and, by extension, absolute ages of sedimentary sequences. Biostratigraphic zonation of the Proterozoic, however, has been frustrated by the extreme morphological conservatism of both prokaryotes and leiosphaerid acritarchs, the signature fossils of the Precambrian record. Distinctive, morphologically complex acritarchs are of relatively recent discovery but promise a much enhanced resolution of at least Neoproterozoic time (Knoll & Butterfield 1989). The rich assortment of ornamented acritarchs in the Svanbergfjellet Formation contributes significantly to this growing record and corroborates and extends various stratigraphic trends seen elsewhere.

One of the most distinctive and widely distributed of these late Proterozoic acritarchs is *Trachyhystrichosphaera aimika*, now known from at least 15 localities worldwide, including the Svanbergfjellet Formation (see systematic section). Without exception, it occurs in rocks of Late Riphean age and appears to be an excellent index fossil for the pre-Vendian Neoproterozoic. In silicified carbonate facies *T. aimika* (= *T. vidalii*) regularly co-occurs with another distinctive acritarch, *Cymatiosphaeroides kullingii* (Knoll & Calder 1983; Knoll 1984; Knoll *et al.* 1991; Allison & Awramik 1989; present study) and a late Riphean age for this association has been corroborated by chemostratigraphy (Kaufman *et al.* 1992). *Trachyhystrichosphaera aimika* also occurs in cherts and shales of the minimally 723±4 Ma old Wynniatt Formation of arctic Canada (Butterfield & Rainbird 1988; Heaman *et al.* 1992), thus providing an absolute-age tiepoint for the taxon. As the Wynniatt and Svanbergfjellet also share comparable species of *Comasphaeridium* and *Germinosphaera*, as well as distinctive $\delta^{13}C$ and $^{87}Sr/^{86}Sr$ geochemical signatures (Asmerom *et al.* 1991), the two sections can be considered broadly correlative and the Svanbergfjellet Formation ca. 700–750 Ma old.

Ornamented acritarchs of pre-Ediacaran age are often very much larger than Paleozoic forms, typically measuring several hundreds (*vs.* tens) of micrometers in diameter (Timofeev *et al.* 1976; Yin & Li 1978; Knoll 1984; Zhang 1984; Yin L. 1985a; Chen & Liu 1986; Zang & Walter 1992a; Knoll & Butterfield 1989; Allison & Awramik 1989; Jankauskas *et al.* 1989; Vidal 1990; Yin 1990; Knoll *et al.* 1991; Jenkins *et al.* 1992; Knoll 1992; Yan & Zhu 1992). Thus, even in the absence of detailed taxonomy, this 'grade of organization' may serve as a coarse biostratigraphic marker for the pre-Ediacaran Neoproterozoic. Smaller, 'Paleozoic-aspect' acritarchs, however, are not exclusively Paleozoic. The Svanbergfjellet assemblage includes species of *Comasphaeridium*, *Dictyotidium*, and *Goniosphaeridium* that would not appear out of place in Cambrian rocks. Their 'unusual' occurrence here is partly explained by the combination of exceptional preservation and the non-disruptive search procedure (i.e. in bedding-parallel thin section); even so, they stand in contrast with their abundant and widely recoverable Paleozoic counterparts. The full biostratigraphic potential of these and other distinctive Proterozoic acritarchs awaits comparable studies elsewhere (e.g., Butterfield & Rainbird 1988), particularly as they may be corroborated by various chronometric (Heaman *et al.* 1992) and chemostratigraphic (Asmerom *et al.* 1991) techniques.

Systematic paleontology

The Svanbergfjellet fossils are classifiable at various taxonomic levels and with varying degrees of confidence. Six broad categories are used in the following systematic treatment: (1) multicellular eukaryotes classified under modern protistan taxonomy (p. 18); (2) multicellular eukaryotes – *incertae sedis* (p. 23); (3) unicellular eukaryotes – *incertae sedis* (=Acritarcha; p. 29); (4) unambiguous cyanobacteria (p. 47); (5) probable cyanobacteria (p. 54); and (6) overall *incertae sedis*, in which all higher-order taxonomy is uncertain (p. 72).

The format of this taxonomy is hierarchical with respect to diagnoses. Thus, a species diagnosis details only those features that distinguish species within a genus, a generic diagnosis those that distinguish genera, etc. Full characterization of the fossil material examined in this study is given in the 'Description' section. With respect to synonymies, the approach taken has been with an eye to excising redundant taxonomic names rather than listing all known occurrences of a particular form. Consequently, most synonymy lists are constructed of only the first descriptions of named taxa. A listing of all junior synonyms and their current placement (as here assessed) is given in the Appendix.

All type and illustrated specimens are housed in the Paleobotanical Collections of the Harvard University Herbaria and assigned Harvard University Paleobotanical Collection (HUPC) numbers. Location and position coordinates for each of the more than 2000 specimens measured in this study are stored with the collections.

Domain Eucarya Woese, Kandler & Wheelis, 1990

Division Chlorophyta Pascher, 1914

Class Chlorophyceae Kützing, 1845

Order Chlorococcales Fritsch, 1935

Genus *Palaeastrum* Butterfield, n.gen.

Type species. – *Palaeastrum dyptocranum* n.sp.

Diagnosis. – Colonial, spheroidal to ellipsoidal cells with prominent intercellular attachment discs; discs circular with a reinforced rim. Colonies monostromatic.

Discussion. – *Palaeastrum* n.gen. differs fundamentally from simple pluricellular aggregates such as *Ostiana* (Fig. 5F–I). The intercellular attachment discs are not simply the product of cell–cell contact but are fully differentiated structures involved in the maintenance of colony structure. In many instances the fossil is represented almost solely by these distinctively arrayed discs (Fig. 5B), the undifferentiated cell walls having considerably less preservation potential.

The attachment discs and mode of colony formation in *Palaeastrum* n.gen. have close morphological analogues among extant chlorococcalean green algae. *Pediastrum* and *Coelastrum*, for example, form multicellular coenobia in which the cells are attached to one another by oriented and differentiated attachment 'plaques'. In *Coelastrum* the plaques are thickened overall but particularly so at their perimeters (Fig. 5D; Marchant 1977), much as in the fossil material. Interestingly, both *Pediastrum* and *Coelastrum* are characterized by walls containing sporopollenin (Atkinson *et al.* 1972; Marchant 1977), a feature that has considerably enhanced the likelihood of their preservation in Phanerozoic rocks (Gray 1960; Evitt 1963a).

Palaeastrum n.gen. colonies are monostromatic and vary from ca. 15 cells (Fig. 5A) to several hundreds of cells, traceable in bedding-parallel thin section for over 1 mm. Unless its colonies were very large and most of the specimens represent transported fragments, it is unlikely that *Palaeastrum* had a determined size (i.e. cell number). By contrast, the coenobia of *Pediastrum* (monostromatic) and *Coelastrum* (spheroidal; Fig. 5D) have a set number of cells. *Palaeastrum* is nevertheless sufficiently comparable to these modern forms to establish its similar grade of multicellular organization and to infer its probable affiliation to the chlorococcalean chlorophytes.

The integrated colonies of *Eosaccharomyces* from the Neoproterozoic of Siberia (Hermann 1979) are broadly comparable to those of *Palaeastrum*. They differ in their lack of differentiated intercellular attachment structures, their distinctive 'streaming' habit, and their larger average cell size. A single Svanbergfjellet 'colony' containing several intimately interconnected cells up to 70 μm across might be assignable to *Eosaccharomyces* (Fig. 5E).

Etymology. – From the Greek *palaios*, ancient, and *astron*, star, with reference to the fossil's antiquity and its morphological comparison of the fossil with extant *Coelastrum* and *Pediastrum*.

Palaeastrum dyptocranum Butterfield, n.sp.
Fig. 5A–C

Holotype. – HUPC 62708, Fig. 5A; Slide 86-G-62-46, England-Finder coordinates M-48-1.

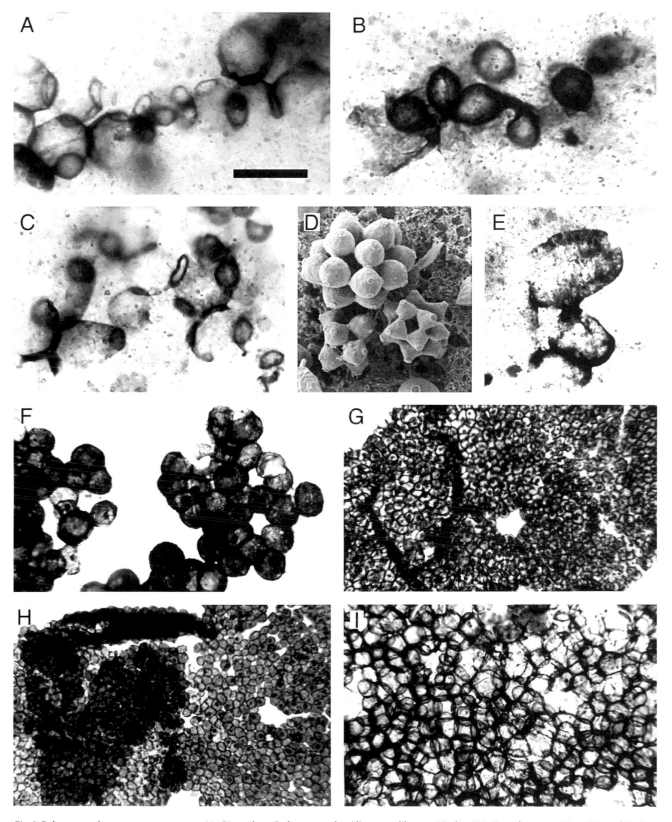

Fig. 5. Palaeastrum dyptocranum n.gen., n.sp. (A–C); modern *Coelastrum proboscidium* var. *dilatatum* Vischer (D); *Eosaccharomyces*(?) sp. (E); and *Ostiana microcystis* Hermann (F–I). From shales of the Algal Dolomite Member, Geerabukta. Scale bar in A equals 20 μm for A–D; 50 μm for E, F, I; 125 μm for G, H. □A. HUPC 62708; 86-G-62-46 (M-48-1); holotype. □B. HUPC 62722; 86-G-62-10 (L-38-3). □C. HUPC 62723; 86-G-61-1 (R-53-2). □D. SEM of modern *Coelastrum* (courtesy of H.J. Marchant; with permission from the *Journal of Phycology*). □E. HUPC 62724; 86-G-62-14 (Q-28-3). □F. HUPC 62725; 86-G-62-190M (O-14-1); loosely aggregated morph. □G. HUPC 62726; 86-G-62-193M (O-28-0); monostromate morph. □H. HUPC 62727; 86-G-62-190M (Q-42-0); monostromate morph. □I. HUPC 62728; 86-G-62-40M (J-36-4); bi-stromate morph.

Type locality. – Algal Dolomite Member, Svanbergfjellet Formation, Geerabukta (79°35'30"N, 17°44"E); 55 m above base of member.

Material. – Seven (7) colonies from shale samples 86-G-62, 86-G-61 and B-2-2. Two (2) designated paratypes: HUPC 62722, Fig. 5B; HUPC 62723, Fig. 5C.

Diagnosis. – A species of *Palaeastrum* with cells 10–25 µm in diameter. Colonies monostromatic.

Description. – Spheroidal to ellipsoidal cells 12–20 µm in diameter ($\bar{x}$ = 16.7 µm; s.d. = 2.0 µm; n = 20) connected to one another by relatively thick-walled attachment discs; each cell with 3–6 (typically 4) attachment discs and accompanying adjacent cells. Resultant colonies monostromatic, comprising tens to hundreds of cells. Attachment discs robust with a reinforced rim; 4–11 µm in diameter ($\bar{x}$ = 6.7 µm; s.d. = 1.9 µm; n = 32). Extracellular layer(s) absent.

Etymology. – From the Greek *dypto* – dive and *kranos* – helmet, with reference to the diving-helmet appearance of individual cells.

Class Ulvophyceae Mattox & Stewart, 1984

Order Siphonocladales Oltmanns, 1904

Genus *Proterocladus* Butterfield, n.gen.

Type species. – *Proterocladus major* n.sp.

Diagnosis. – Multicellular, uniseriate and occasionally branched filaments with intercellular septa. Cells thin-walled, psilate and cylindrical; length highly variable but typically much longer than wide. Branches usually subjacent to a septum in the primary axis and themselves often septate. Apical terminations simply rounded or capitate.

Discussion. – *Proterocladus* n.gen. has close morphological analogues among the extant Chlorophyta, in particular with the branching septate filaments of *Cladophoropsis* and *Cladophora* (Fig. 7D). Some simple Rhodophyta have a broadly comparable morphology (e.g., *Rhodochorton*); however, the marked variance in *Proterocladus* cell length is especially reminiscent of that observed in modern *Cladophoropsis*, while its branching patterns and septal structure are comparable to those of *Cladophora* (these two extant chlorophytes are closely related members of the Siphonocladales; p. 14).

The relatively robust, two-dimensional septa of *Proterocladus* n.gen. show less variation in diameter than the thin-walled and often expanded cells and are therefore used as the principal measure of filament width. Broadly discrete size ranges and accessory qualitative differences warrant the recognition of three species, *P. major* n.sp., *P. minor* n.sp., and *P. hermannae* n.sp.

Etymology. – From the Greek *proteros* – earlier and *klados* – branch, with reference to its branching habit and similarity to modern *Cladophora*.

Proterocladus major Butterfield, n.sp.
Figs. 6A–J, 7C

Holotype. – HUPC 62709, Fig. 6C; Slide 86-G-62-53, England-Finder coordinates L-11-2.

Type locality. – Algal Dolomite Member, Svanbergfjellet Formation, Geerabukta (79°35'30"N, 17°44'E); 55 m above base of member.

Material. – One hundred sixty two (162) cells measured from ca. 50 thallus fragments. From shale sample 86-G-62. Ten (10) designated paratypes: HUPC 62730, Fig. 7C; HUPC 62732–62740, Fig. 6A–B, D–J.

Diagnosis. – A species of *Proterocladus* with robust septa 10–35 µm in diameter. Filaments typically constricted at septa.

Description. – Multicellular, uniseriate and occasionally branched filaments. Cells cylindrical, 13–52 µm wide ($\bar{x}$ = 32 µm; s.d. = 8 µm; n = 162) and 52–731 µm long ($\bar{x}$ = 322 µm; s.d. = 143 µm; n = 77; if specimens with only one septum are included the maximum cell length exceeds 990 µm [$\bar{x} \geq 320$ µm; s.d. ≥ 158 µm; n = 153]). Septa circular and robust, 10–32 µm in diameter ($\bar{x}$ = 20; s.d. = 4; n = 158). Branches usually subjacent to a septum in the primary axis and sometimes fully isolated by a second axial septum; angle of divergence 45–90°; one to several branches per cell. Average branch diameter ca. 80% that of the 'primary' axis. Cell walls psilate but with a characteristic micro-fractured texture. Apical terminations rounded and occasionally capitate.

Discussion. – *Proterocladus major* n.sp. typically occurs as transported fragments of a few to ca. 20 cells but whole thalli were sure to have formed conspicuous tufts or algal carpets; one specimen (Fig. 6G) extends for nearly 1 cm in bedding-parallel thin section. Another shows what appear to be six separate filaments radiating from the apex of a single cell (Fig. 6J), indicating a considerable capacity for dense multicellular growth.

The septa of *P. major* n.sp. are occasionally preserved in parallel-to-bedding orientation where they can be confirmed as being circular (Fig. 6A, I). In one instance a septum appears to be perforated by a central pore (Fig. 6A) inviting comparison with the primary pit connections of red algae, or with the enigmatic chlorophyte *Smithsoniella* Sears & Brawley, 1982. Another specimen, however, shows no such structure (Fig. 6I), and the pore may be simply a reflection of a centripetal septum formation such as occurs in *Cladophora*, or a taphonomic feature.

Etymology. – With reference to its large size.

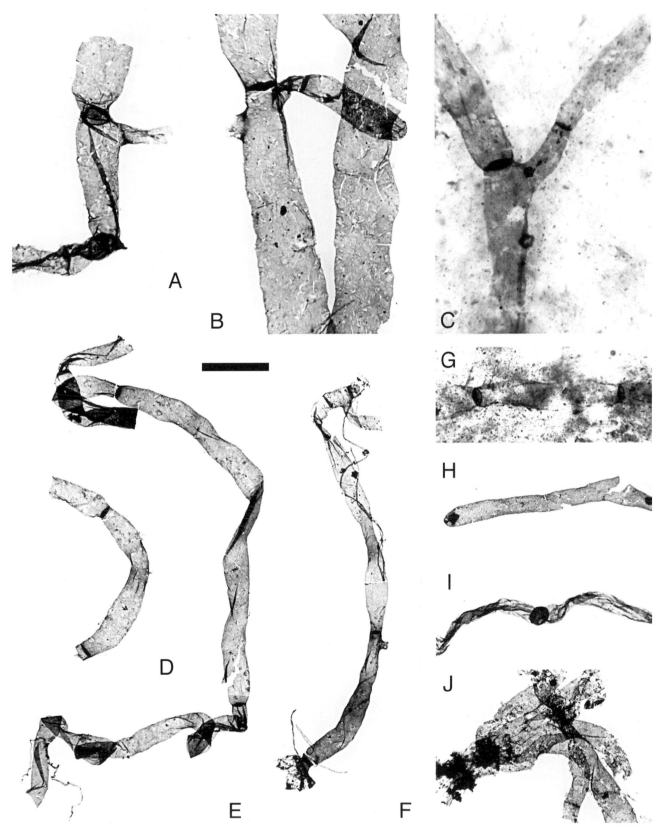

Fig. 6. Proterocladus major n.gen., n.sp. From shales of the Algal Dolomite Member, Geerabukta. Scale bar in E equals 45 μm for A–C; 60 μm for I; 100 μm for D–H, J. □A. HUPC 62732; 86-G-62-22M (Q-33-1). □B. HUPC 62733; 86-G-62-18M (G-37-0). □C. HUPC 62709; 86-G-62-53 (L-11-2); holotype; in bedding-parallel thin-section. □D. HUPC 62734; 86-G-62-17M (S-36-0). □E. HUPC 62735; 86-G-62-90M (G-31-0). □F. HUPC 62736; 86-G-62-76M (O-25-0). □G. HUPC 62737; 86-G-62-10 (P-25-0). □H. HUPC 62738; 86-G-62-19M (U-36-3); terminal cell. □I. HUPC 62739; 86-G-62-87M (L-27-0); collapsed filament with septum. □J. HUPC 62740; 86-G-62-79M (O-37-1); single cell with plexus of 5 branches.

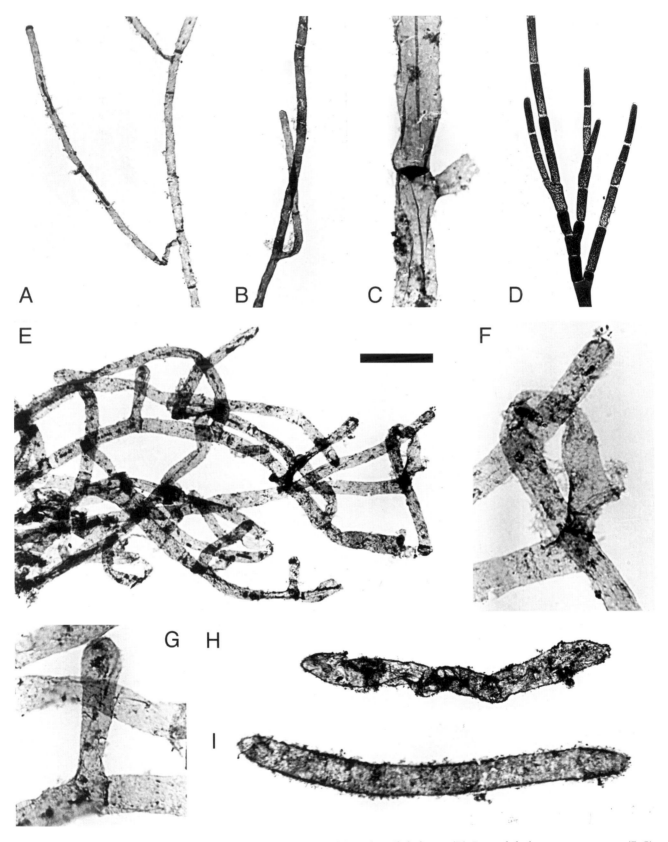

Fig 7. Proterocladus minor n.gen., n.sp. (A, B); *Proterocladus major* n.gen., n.sp. (C); modern *Cladophora* sp. (D); *Proterocladus hermannae* n.gen., n.sp. (E–G); and *Digitus adumbratus* n.sp. (H, I). All fossils from shales of the Algal Dolomite Member, Geerabukta. Scale bar in E equals 50 μm for A–C, E; 185 μm for D; 20 μm for F–G; 75 μm for H–I. □A. HUPC 62710; 86-G-62-15M (N-24-0); holotype. □B. HUPC 62729; 86-G-62-14M (L-40-2). □C. HUPC 62730; 86-G-62-16M (L-41-4). □D. *Cladophora* sp. □E. HUPC 62711; 86-G-62-93M (P-44-0); holotype. □F. HUPC 62711; detail of E; ?zoospore release structure. □G. HUPC 62711; detail of E; branch and septum. □H. HUPC 62731; 86-G-62-119M (G-36-1). □I. HUPC 62719; 86-G-62-133M (N-29-2); holotype.

Proterocladus minor Butterfield, n.sp.

Fig. 7A–B

Holotype. – HUPC 62710, Fig. 7A; Slide 86-G-62-15M, England-Finder coordinates N-24-0.

Type locality. – Algal Dolomite Member, Svanbergfjellet Formation, Geerabukta (79°35'30"N, 17°44'E); 55 m above base of member.

Material. – Twenty (20) cells measured from 5 thalli. From shale sample 86-G-62. One (1) designated paratype: HUPC 62729, Fig. 7B.

Diagnosis. – A species of *Proterocladus* with robust septa 3–7 µm in diameter. Filaments rarely constricted at septa.

Description. – Multicellular, uniseriate, occasionally branched filaments. Cells cylindrical, 5–12 µm wide ($\bar{x} = 6.8$ µm; s.d. $= 1.5$ µm; $n = 20$) and 30–160 µm long ($\bar{x} = 91$ µm; s.d. $= 40$ µm; $n = 12$; if specimens with only one septum are included the maximum cell length exceeds 175 µm [$\bar{x} \geq 94$ µm; s.d. ≥ 41 µm; $n = 20$]). Septa robust, 3–7 µm in diameter ($\bar{x} = 5.4$ µm; s.d. $= 1.0$ µm; $n = 19$). Branches subjacent to a septum in the primary axis and usually isolated by a second axial septum; angle of divergence ca. 90°. Branches approximately the same diameter as the primary axis. Cell contents sometimes preserved as dark elongate inclusions. Apical terminations rounded.

Discussion. – In addition to having smaller dimensions, *Proterocladus minor* n.sp. is distinguished from *P. major* n.sp. by an absence of significant constrictions at septa and the approximately equal diameters of branches and primary axes. Like *P. hermannae* n.sp., *P. minor* occasionally preserves condensed cell contents but is distinguished by its smaller size and its more robust and relatively numerous septa.

Etymology. – With reference to its small size relative to the type species.

Proterocladus hermannae Butterfield, n.sp.

Fig. 7E–G

Holotype. – HUPC 62711, Fig. 7E–G; Slide 86-G-62-93M, England-Finder coordinates P-44-0.

Type locality. – Algal Dolomite Member, Svanbergfjellet Formation, Geerabukta (79°35'30"N, 17°44'E); 55 m above base of member.

Material. – One (1) multicellular thallus from shale sample 86-G-62.

Diagnosis. – A species of *Proterocladus* with insubstantial septa and filaments 7–14 µm in diameter.

Description. – Multicellular, uniseriate and occasionally branched filaments, 7–14 µm in diameter, forming extensive thallus. Cells much longer than wide. Septa uncommon and insubstantial, usually associated with branches. Maximally one branch per cell, which diverges at right angles from the parental axis. Branches with a slight basal constriction but otherwise communicating freely with the parental cell; angle of divergence ca. 90°. Rare, conical protrusions in the lateral wall. Cell contents sometimes preserved as dark rod-shaped inclusions. Apical terminations simply rounded.

Discussion. – The pattern in *P. hermannae* n.sp. of large but irregularly sized cells and branches associated with a single suprajacent septum is indistinguishable from that found in the modern siphonocladalean green alga *Cladophoropsis*. Moreover, the single conical protrusion (Fig. 7F), which is not associated with a septum, is identical to the zoospore release structures of living *Cladophoropsis* (Børgesen 1913, p. 48, Fig. 33; O'Kelly & Floyd 1984). Together these features are sufficiently distinctive to allow identification of the cell-division program of *P. hermannae* ('segregative cell division'; p. 14) and thereby of its original pigment complement, taxonomic affiliation and overall biology.

Etymology. – In honor of Tamara N. Hermann for her pioneering studies in Proterozoic paleontology.

Incertae sedis

Genus *Pseudotawuia* Butterfield, n.gen.

Type species. – *Pseudotawuia birenifera* n.sp.

Diagnosis. – Thin-walled tomaculate macrofossil with a terminal pair of dark reniform structures.

Etymology. – With respect to its superficial resemblance to *Tawuia*.

Pseudotawuia birenifera Butterfield, n.sp.

Fig. 8A

Holotype. – HUPC 62741, Fig. 8A; shale specimen 86-G-30-3BP.

Type locality. – Algal Dolomite Member, Svanbergfjellet Formation, Geerabukta (79°35'30"N, 17°44'E); 38 m from base of member.

Material. – One (1) specimen from shale sample 86-G-30.

Diagnosis. – A species of *Pseudotawuia* ca. 2 mm wide and ca. 1 cm long.

Description. – Thin-walled organic film, 1.84 mm wide and ≥9.45 mm long, bearing a bilaterally(?) symmetrical pair of dark, ca. 0.75 mm long, reniform structures at one end.

Discussion. – *Pseudotawuia birenifera* n.sp. differs conspicuously from *Tawuia dalensis* Hofmann, 1979. It is defined by

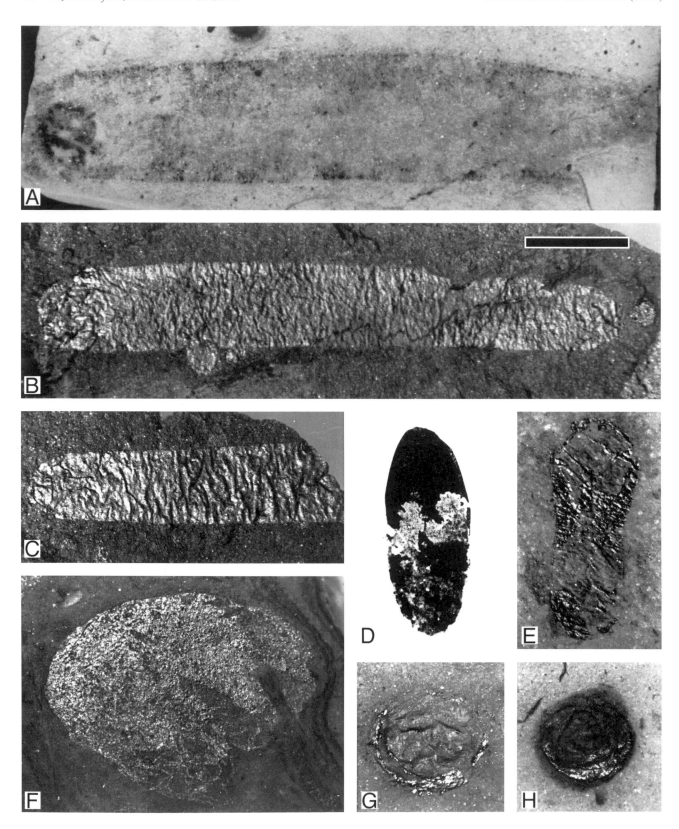

Fig. 8. Pseudotawuia bireniformis n.gen., n.sp. (A); *Tawuia dalensis* Hofmann (B–E); large spheroidal compression (F); and *Chuaria circularis* Walcott (G, H). From shales of the Algal Dolomite Member, Geerabukta. Scale bar in A equals 1.5 mm. □A. HUPC 62741; 86-G-30-3BP; holotype. □B. HUPC 62742; 86-G-62-1BP; with associated *Chuaria* sp. and *Cerebrosphaera buickii* n.sp. □C. HUPC 62743; 86-G-62-3S; now isolated and mounted for SEM. □D. HUPC 62744; 86-G-62-108M (K-32-0); isolated specimen. □E. HUPC 62745; 86-G-61-1BP; '*Pumilabaxa*' morph. □F. HUPC 62746; 86-G-61-3BP. □G. HUPC 62747; 86-G-61-2BP. □H. HUPC 62748; 86-G-28-2BP; with associated 'organic stain'.

a thin organic film that has left no sedimentary imprint (co-occurring *Cerebrosphaera* n.gen. leave prominent imprints) and, at one end, bears a symmetrical pair of dark reniform structures. These are unrelated to the mechanically induced 'disc-like terminal structures' on some of the Little Dal *Tawuia* (cf. Hofmann 1985a), and, if they are indeed bilaterally symmetrical, represent a grade of organization generally thought to be limited to Ediacaran and younger organisms. The combination of large size, thin-walled cuticle, elongate shape, and paired terminal structures in *P. birenifera* invites comparison with simple vermiform metazoa. The absence of cellularity in the main body is not in itself evidence against metazoan multicellularity, since animal cells lack walls and are not expected to preserve as organic-walled fossils (Butterfield 1990); the remains in such cases would be cuticular. In any event, the paired terminal structures point to significant cellular differentiation in *Pseudotawuia* and, unless these are unusually large individual cells, would appear to derive from differentiated, multicellular tissues or organs.

Etymology. – From the Latin *bi*, two, *ren*, kidney, and *fero*, bear, with reference to the paired reniform structures.

Genus *Tawuia* Hofmann, 1979

Synonymy. – ☐1979 *Tawuia* Hofmann, n.g. – Hofmann & Aitken, pp. 157–158. ☐1980 *Ellipsophysa* Zheng (gen. nov.) – Zheng, p. 60. ☐1980 *Pumilibaxa* Zheng (gen. nov.) – Zheng, p. 61. ☐1980 *Nephroformia* Zheng (gen. nov.) – Zheng, p. 62. ☐1985 *Glossophyton* Duan et Du (gen. nov.) – Xing *et al.*, p. 72. ☐1989 *Mesonactus* gen. nov. – Fu, pp. 74, 77. ☐1989 *Tachymacrus* gen. nov. – Fu, pp. 74, 77. ☐1992 *Tawuia dalensis* Hofmann, 1979 – Zang & Walter 1992bs, pp. 312–313, Pl. 4A, D–I (see for extended synonymy). ☐1993 *Luonanconcha* Jian et Hu gen. nov. – Hu *et al.*, pp. 100, 106.

Type species. – *Tawuia dalensis* Hofmann, 1979, pp. 158–160.

Discussion. – Most thick-walled, more or less tomaculate (sausage-shaped) Proterozoic macrofossils can be reliably assigned to the form genus *Tawuia*. The principal difficulty in its identification lies in an apparent morphological gradation into ellipsoidal and circular forms. Thus, while Hofmann (1985a, b, 1992) considers genera such as *Shouhsienia* and *Ellipsophysa* to be short representatives of *Tawuia*, Zhang R. *et al.* (1991) prefer to retain these ellipsoidal fossils as a distinct form. It is worth noting in this regard that the distortion of a spheroid under compression will often yield an ellipsoid with a length:width ratio of ca. 1.5 (Harris 1974, Text-figs. 5.C–H, 6), or considerably more if the vesicle is split (Harris 1974, p. 139). Unlike *Tawuia*, *Shouhsienia* tends to have a single terminal split, suggesting that much of its aspect ratio is taphonomically induced and that its taxonomic affiliation is with the spheroidal acritarchs, e.g., *Leiosphaeridia wimanii* n.comb. (Fig. 13E–F). On the other hand,

unsplit ellipsoids with a length:width ratio as small as 2.5 (Fig. 8D–E) can, on independent structural grounds, be reliably assigned to *Tawuia* (see below).

A single species of *Tawuia* is recognized by Hofmann (1985a); others may exist but they await rigorous documentation. Like *T. siniensis* Duan, 1982, *Tawuia* in the Svanbergfjellet Formation are markedly smaller and less curved than the mean/mode of the Little Dal type material; however, the size range and habit of both these populations fall entirely within those of *T. dalensis* (Hofmann 1985a), thus leaving the putative second species without diagnostic (i.e. unique) characters. In view of the relatively small numbers of Svanbergfjellet *Tawuia* and the difficulty in assessing discrete species from the literature, only its genus-level synonymy is considered here. An extensive (to 1988) collation of Proterozoic carbonaceous megafossils is given by Hofmann (1992).

Tawuia dalensis Hofmann, 1979

Figs. 8B–E, 23H

Material. – Sixteen (16) specimens (including 5 counterparts) from shale samples 86-G-62 and 86-G-61; 4 isolated by acid maceration (2 on glass slides, 2 on SEM stubs).

Description. – Macroscopic, carbonaceous, tomaculate fossils with more or less parallel sides and rounded ends. Complete length 3.07–8.57 mm ($\bar{x} = 4.96$ mm; $n = 3$); width 0.94–1.90 mm ($\bar{x} = 1.39$ mm; s.d. = 0.30 mm; $n = 11$); length:width ratio 2.5–6.0 ($\bar{x} = 3.7$; $n = 3$); occasionally somewhat narrower medially, and/or towards one end. Wall bi-layered: 'primary' outer(?) layer opaque, brittle, and minimally 1 μm thick; 'secondary' inner(?) layer translucent, flexible, and ca. 3.5 μm thick (possibly 3.5/2 = 1.75 μm thick). Walls typically with transverse wrinkles. Primary splits absent.

Discussion. – The carbonaceous walls of Svanbergfjellet *T. dalensis* are very well preserved and can be freed from their shale matrix by careful acid maceration. SEM (Fig. 23H) and transmitted light microscopy (Fig. 8D) of a number of these isolated specimens reveals a compound wall structure: a 1 μm thick, brittle and opaque wall is lightly bonded to a 3.5 μm thick but remarkably flexible translucent layer. This translucent layer appears to have been internal to the dark one (it was not found on the back side of the specimen in Fig. 8C after acid dissolution of the shale matrix) and may thus represent the two fused sides of a collapsed inner cylinder; it nevertheless comprises the bulk of a *Tawuia* fossil and is capable of maintaining its form, even in the absence of the more conspicuous opaque wall (Fig. 8D). The substantial overall thickness of *T. dalensis* (minimally 5.5 μm) is comparable to that of *Chuaria circularis* and undoubtedly contributes to its similar occurrence as imprints and compressions in diverse sediments.

The observation of two discrete wall layers provides significant new detail in delineating *T. dalensis* morphology, but little towards a resolution of its overall biology. If, however, the outer wall can be shown to be fundamentally opaque (p. 12), then it would be unlikely that *T. dalensis* carried on a photosynthetic metabolism, at least in this particular phase of its life cycle. Its large size and complex histology suggests that it was a true multicellular, possibly coenocytic, organism; otherwise, its taxonomic affiliations remain obscure.

Svanbergfjellet *T. dalensis* are commonly associated with large discoidal (spheroidal) fossils. Some of these are possibly *Chuaria* (although not *C. circularis*), but many are clearly *Cerebrosphaera* n.gen.; for example, the ca. 420 μm diameter fossil just off the end of the *T. dalensis* specimen in Fig. 8B. Whether these co-occurrences represent a meaningful association (i.e. a '*Chuaria-Tawuia* assemblage') is uncertain because of small sample size; *Cerebrosphaera* is certainly far more widespread than *Tawuia* in the Svanbergfjellet shales.

Genus *Valkyria* Butterfield, n.gen.

Type species. – Valkyria borealis n.sp.

Diagnosis. – Complex, thin-walled carbonaceous fossils: tomaculate main body with medially borne, lobate, and occasionally branched lateral extensions. Main body commonly contains one or more dark circum-terminal discoids and a vaguely defined medial stripe. A centrally-positioned sac-like structure (approximately the same diameter as the main body) and a sub-terminal partition in the main body are occasionally present. Multiple specimens may diverge radially from a large, dark, central body.

Discussion. – Given the diversity and regular integration of its constituent parts it is clear that *Valkyria* n.gen. was a complex, multicellular organism; however, a survey of possible modern analogues yields no convincing *bauplan*-level comparisons. The least sensational, though not obviously correct, interpretation would ally it to some of the complex algae that develop both axial filaments of 'unlimited' growth and lateral filaments of limited or determinate growth: e.g., *Batrachospermum* (Rhodophyta), *Draparnaldia* and *Draparnaldiopsis* (Chlorophyta, Chaetophorales), or *Dasycladus* and *Batophora* (Chlorophyta, Dasycladales). These modern forms are nevertheless distinct in that they are permanently attached to a substrate, have their lateral axes regularly arranged in whorls on the main axis, and bear their reproductive structures in or on the lateral axes. They furthermore lack structures that would preserve as a medial stripe or the various vesicular constituents of *Valkyria*, and they are all fundamentally larger than the fossil. Species of *Valonia* (Siphonocladales) satisfy some of these concerns (e.g., smaller size, irregular axes), but otherwise offer no more tenable a comparison.

Etymology. – After the Valkyries of Norse mythology; the armed warrior-maidens of Odin, choosers of the slain.

Valkyria borealis Butterfield, n.sp.

Figs. 9A–E, 10A–H, 11

Holotype. – HUPC 62712, Fig. 9A; Slide 86-G-62-5M, England-Finder coordinates L-29-1.

Type locality. – Algal Dolomite Member, Svanbergfjellet Formation, Geerabukta (79°35'30"N, 17°44'E); 55 m above base of member.

Material. – Eighty (80) specimens isolated by acid maceration and mounted on glass slides. Five (5) additional specimens and a 'colony' consisting of ca. 12 individuals identified in bedding-parallel thin section. From shale sample 86-G-62. Thirteen (13) designated paratypes: HUPC 62749-62760, Figs. 9B–E, 10E–H; HUPC 62907, Fig. 11.

Diagnosis. – A species of *Valkyria* with the central body 100–1000 μm long and 20–200 μm wide.

Description. – Thin-walled, carbonaceous, tomaculate (sausage-shaped) bodies 164–930 μm long ($\bar{x}$ = 465 μm; s.d. = 200 μm; n = 25) and 25–182 μm wide ($\bar{x}$ = 95 μm; s.d. = 29 μm; n = 79) that bear 1–14 lateral extensions; entire specimens with a mean aspect ratio of 5.5:1 (s.d. = 1.2:1; n = 25). Lateral axes occur only on the central portion of the main body, never the ends, with the attachment points marked by prominent circular scars, 5–34 μm in diameter ($\bar{x}$ = 17 μm; s.d. = 5 μm; n = 78). Lateral axes typically of a lobate construction; hollow and apparently non-septate; uniramous or branched. The main body commonly bears one to several circum-terminal dark circular bodies (present in 41 of 85 specimens) ca. 30 μm in diameter, and a more or less vaguely defined longitudinal medial stripe (present in 24 specimens). A few specimens have an inclusion-rich terminal portion (ca. one-third) of their main body separated by a prominent partition. Three specimens bear a large, centrally-positioned, thick-walled vesicle and associated structures that occupy most of the diameter of the main body. In one instance, three 'whole' specimens of *V. borealis* n.sp. diverge radially from a large (ca. 450 μm in diameter), ill-defined opaque body. A number of specimens preserve entangled filaments of *Siphonophycus septatum* within the main body.

Discussion. – Valkyria borealis n.sp. preserves at least six distinct cell (or tissue) types: (1) the thin-walled, often lightly shagrinate main body; (2) hollow, uniramous or branched (Fig. 10A, D, F, G) lateral axes, typically of a lobate construction (Fig 9C–E) but occasionally more straight-walled (Fig. 10D, G); (3) dark to opaque circular bodies (spheroids?) that occur near one or both ends of the main body (Figs. 9A, C–E, 10B–C); (4) a longitudinal stripe within the main body which sometimes appears to terminate at one of the dark, circum-terminal discs (Figs. 9A–B, 10C); (5) a large, centrally-positioned, thick-walled vesicle that may occupy a substantial proportion of the main body; one end of this vesicle appears to differentiate a narrow finger-like projection (Fig. 10A, E) and the other a curved discoidal(?) struc-

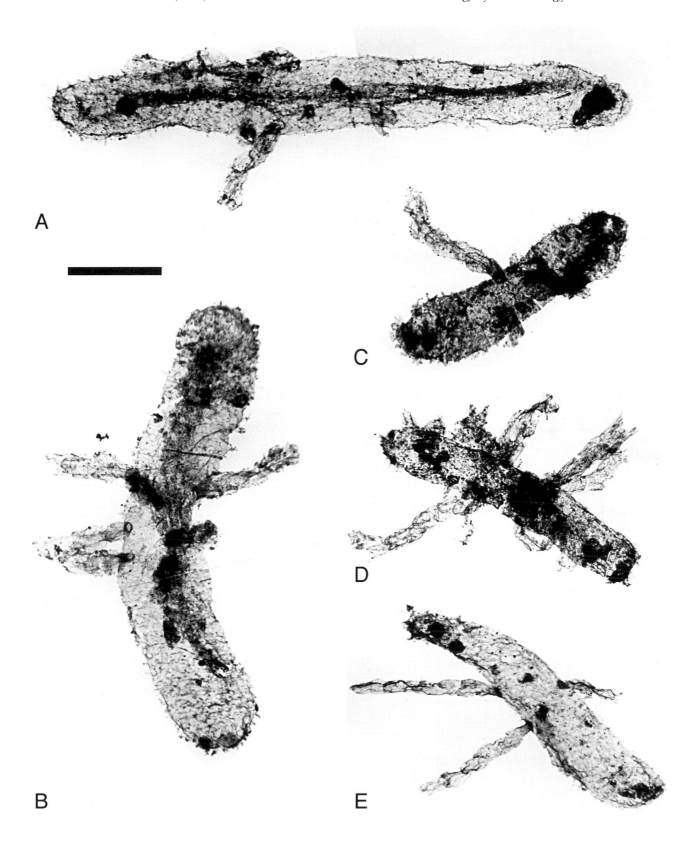

Fig. 9. Valkyria borealis n.gen., n.sp. From shales of the Algal Dolomite Member, Geerabukta. Scale bar in B equals 150 µm. ☐A. HUPC 62712; 86-G-62-5M (L-29-1); holotype. ☐B. HUPC 62749; 86-G-62-7M (M-29-4). ☐C. HUPC 62750; 86-G-62-6M (R-29-4). ☐D. HUPC 62751; 86-G-62-2M (P-36-0). ☐E. HUPC 62752; 86-G-62-58M (L-32-0).

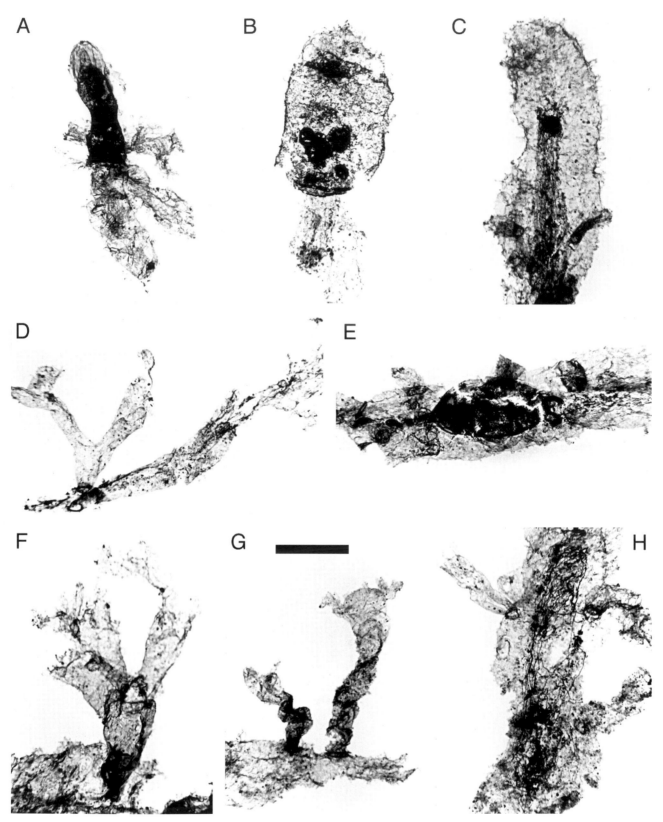

Fig. 10. Valkyria borealis n.gen., n.sp. From shales of the Algal Dolomite Member, Geerabukta. Scale bar in G equals 50 μm for A, D, F; 80 μm for B, E, G, H; 125 μm for C. □A. HUPC 62753; 86-G-62-236M (M-41-0); with central vesicle showing a pointed terminal structure and a possible orifice in the adjacent outer wall. □B. HUPC 62754; 86-G-62-73M (N-21-3); with 'partitioned' terminal segment. □C. HUPC 62755; 86-G-62-64M (P-41-0). □D. HUPC 62756; 86-G-62-41M (Q-19-1); branched, lateral axes. □E. HUPC 62757; 86-G-62-10M (N-29-3); with central vesicle, one end pointed, the other with a curled structure. □F. HUPC 62758; 86-G-62-4M (N-31-4); branched lateral axis. □G. HUPC 62759; 86-G-62-58M (K-40-4); branched lateral axis. □H. HUPC 62760; 86-G-62-73M (S-29-2); specimen in which the cavity of the main body has been occupied by filamentous micro-organisms.

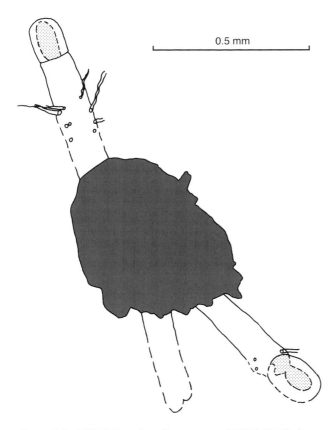

Fig. 11. Colonial(?) *Valkyria borealis* n.gen, n.sp. (IIUPC 62907); drawn from shale thin-section 86-G-62-35 (H-52-0); Algal Dolomite Member, Geerabukta. Note that the partitioned distal segments of the two principal units are the same as the isolated structure in Fig. 10B.

ture (Fig. 10E); and (6) a large, poorly defined opaque ovoid with a radial texture to which several *V. borealis* 'individuals' may be attached at one end (Fig. 11). These six types of course represent only those 'cell types' that were both preservable and physically contiguous with recognizable *Valkyria* fossils; the true number must have been higher (e.g., unwalled or acuticular cells/tissues, gametes, alternate generations). Most *V. borealis* n.sp. are 'entire' and lack obvious attachment structures, thus suggesting a free-floating, planktic existence. On the other hand, an occurrence of ca. 12 specimens localized in ca. 25 mm² of a single bedding-parallel thin section points to their at least occasionally gregarious habit. Three of these specimens are attached at one end to a large, poorly defined opaque ovoid (Fig. 11), which conceivably represents a germinating or developing reproductive body. This gregarious population is further characterized by a prevalence of individuals in which the distal (in the case of the attached specimens) ca. one-third of the main body is separated by a prominent crosswall and contains elevated amounts of condensed organic matter (Figs. 10B, 11).

Fifteen of the 85 measured specimens of *V. borealis* n.sp. contain substantial accumulations of *Siphonophycus sep-*

tatum filaments within the main body and, in one instance, appear to extend into a lateral axis (Fig. 10H). Their occurrence here is clearly not just happenstance and they possibly represent degrading heterotrophic microbes. Alternatively, the filaments may have been symbiotic within (or parasitic upon) the larger organism.

Etymology. – From the Latin *borealis* – northern, with reference to the high latitude of the fossil occurrence.

Group Acritarcha Evitt, 1963

Genus *Cerebrosphaera* Butterfield, n.gen.

Type species. – *Cerebrosphaera buickii* n.sp.

Diagnosis. – Spheroidal vesicles with regularly and prominently wrinkled walls. Wrinkles sinuous: anastomosing, interfingering or, rarely, sub-parallel, but never intersecting. Vesicle walls ca. 1.5 µm thick, inelastic, and often opaque. Outer, thin-walled envelope sometimes present.

Discussion. – The diagnostic wrinkling of *Cerebrosphaera* n.gen. walls is possibly taphonomic, but if so, it is derived from a unique underlying wall construction. Its widespread occurrence in Svanbergfjellet shales and common co-occurrence with typical (i.e. unwrinkled) leiosphaerids rule out localized taphonomy as an explanation for its texture. *Cerebrosphaera* is further distinguished from leiosphaerids by an enveloping sheath and its thick, inelastic walls. This latter quality is expressed in its lack of secondary folds and a proclivity for radial fracture under sedimentary compaction (Fig. 12C–F; cf. Harris 1974: type 2 failure). The cerebroid wrinkling of *Cerebrosphaera* walls clearly preceded their fracture, suggesting that the wrinkles may in fact have been primary.

Small (15–150 µm diameter), radially split, and usually opaque spheroidal microfossils with a rough surface have previously been classified as *Turuchanica* Rudavskaja. The recent subsumption of *Turuchanica* into *Leiosphaeridia* by Jankauskas *et al.* (1989) and, more importantly, the absence of regular wrinkling in *Turuchanica*, militates against its comparison with *Cerebrosphaera* n.gen. More reminiscent is a population of thick-walled and occasionally ensheathed spheroids, 80–500 µm in diameter, described as *Chuaria globosa* Ogurtsova & Sergeev, 1989, from the Late Riphean of southern Kazakhstan. These silicified fossils are difficult to compare with the Svanbergfjellet compressions, but they appear to have similarly convoluted walls; their three-dimensional preservation further suggests that the wrinkles were primary or a very early diagenetic feature.

Etymology. – From the Latin *cerebrum*, brain, and *sphaera*, ball, with reference to the cerebroid convolutions of its walls.

Cerebrosphaera buickii Butterfield, n.sp.
Fig. 12A–H

Synonymy. – ☐1991 *Leiosphaeridia* sp. cf. *L. atava* (Naumova) Jankauskas, 1989 – Knoll *et al.*, 1991, p. 558, Fig. 21.2–21.3.

Holotype. – HUPC 62713, Fig. 12D–E; Slide P-2945-47M, England-Finder coordinates S-33-4.

Type locality. – Lower Dolomite Member, Svanbergfjellet Formation, Polarisbreen (79°10'N, 18°12'E); 17 m below the base of the *Minjaria* biostrome.

Material. – One hundred twenty (120) specimens: 12 in bedding-parallel thin section; 6 as exposed bedding-plane compressions; 102 isolated by acid maceration (84 on glass slides, 18 on SEM stubs). From shale samples P-2945, 86-G-33, 86-G-62, 86-G-61, 86-G-28 and 86-G-30. Six (6) designated paratypes: HUPC 62761–62763, Fig. 12A–C; HUPC 62764–62766, Fig. 12F–G.

Diagnosis. – A species of *Cerebrosphaera* with vesicles 100–1000 µm in diameter.

Description. – Spheroidal carbonaceous microfossils, 100–960 µm in diameter ($\bar{x} = 365$ µm, s.d. = 151 µm; $n = 115$), with the primary wall prominently and regularly convoluted. Wrinkles sinuous; anastomosing, interfingering, or rarely sub-parallel, but never intersecting. Psilate walls ca. 1.5 µm thick, often opaque, and relatively inelastic; radial fractures common; secondary folds absent. Thin-walled, enveloping sheath of leiosphaerid type occasionally present.

Discussion. – *Cerebrosphaera buickii* n.sp. is conspicuous in most of the fossiliferous Svanbergfjellet shales, as well as in the immediately overlying Draken Formation (Knoll *et al.* 1991). Its distinctive wrinkling pattern allows the positive identification even of isolated fragments (Fig. 12B). Reflected light (Fig. 12A, E) or SEM (Fig. 12C, G, H) may be necessary to recognize opaque material, while transmitted light reveals the nature of the thin, structureless envelope (Fig 12D). The primary wall of *C. buickii* may be translucent when isolated in single layers (Fig. 12B), but whole specimens are typically opaque, as the two sides of the spheroid double the overall thickness (Fig. 12D, F). Under SEM the substantial thickness of *C. buickii* walls can be measured directly from fractured specimens (Fig. 12H), or by halving the width of the tightest wrinkles (Fig. 12G); both methods indicate a single-wall thickness of ca. 1.5 µm. By contrast, the walls of *Chuaria circularis* are 2–3 µm thick yet were originally much more pliant (i.e. non-fracturing). The SEM of the fractured *C. buickii* specimen in Fig. 12H further reveals numerous rounded perforations 2–3 µm in diameter, possibly the result of heterotrophic activity.

Etymology. – In recognition of the unrelenting advice and skepticism of Dr. Roger Buick, who first identified these fossils as pyrite framboids. He subsequently suggested the generic name.

Genus *Chuaria* Walcott, 1899

Type species. – *Chuaria circularis*, Walcott, 1899, pp. 234–235.

Synonymy. – ☐1899 *Chuaria circularis*, nov. g. and sp. – Walcott, pp. 234–235. ☐1935 *Fermoria*, gen. nov. (Chapman, 1934) – Chapman, pp. 114–115. ☐1935 *Protobolella*, gen. nov. – Chapman, p. 117. ☐1993 *Xunjiansiella* Hua et Jian, gen. nov. – Hu *et al.*, pp. 99, 103–104. ☐*non* 1941 *Chuaria wimani* n.nom. – Brotzen, pp. 258–259 (= *Leiosphaeridia*). ☐*nec* 1989 *Chuaria globosa* Ogurtsova et Sergeev, sp. nov. – Ogurtsova & Sergeev, p. 121 (= ?*Cerebrosphaera*).

Diagnosis. – Thick-walled (single wall >2 µm thick) spheroidal vesicles up to 5000 µm in diameter. Walls psilate, opaque. Radial splits absent. Envelope absent.

Discussion. – *Chuaria* has been widely discussed and variously interpreted, major reviews being offered by Ford & Breed (1973) and Hofmann (1985a, b). Duan (1982) and Sun (1987a) notwithstanding, the consensus view of *Chuaria* is of a hollow, organic-walled spheroid of unknown but probably algal affinities, i.e. an acritarch. This interpretation is supported by the Svanbergfjellet material. An unambiguous *C. circularis* specimen (Fig. 13G [SEM], H [reflected light]) reveals in cross-section (Fig. 13I) the two well-defined sides of a compressed vesicle separated by an internal space.

In its current usage *Chuaria* has come to serve simply as a (form) taxonomic pigeonhole for all relatively large, spheroidal acritarchs (Jankauskas *et al.* 1989). While such application may be useful for a broad characterization of Proterozoic fossil assemblages, it is clear that a disparate array of natural taxa are unnecessarily subsumed into the genus. Large (ca. 1 mm) spheroidal acritarchs in the Svanbergfjellet Formation can be classified into at least three readily distinguishable genera: *Leiosphaeridia*, *Cerebrosphaera* and *Chuaria*. Only the latter two have walls substantial enough to leave a significant sedimentary imprint. *Chuaria* is further distinguished from *Cerebrosphaera* by its thicker, entirely opaque and more pliant walls, simple concentric folding, and absence of an enveloping sheath.

Perhaps the most reliable taxonomic feature of *Chuaria* is its substantial wall thickness (Amard 1992). Unfortunately, this character is usually given in relative terms or, if quantified, the method of measurement left unstated. Electron microscopy of well preserved, extractable material permits its direct and unambiguous measurement. Such examination has previously indicated a wall thickness of 2–2.5 µm (Jux 1977; material from the Chuar type section) and 7–19 µm (Amard 1992; material from west Africa). In the Svanbergfjellet Formation, the 770 µm *Chuaria* specimen mentioned above (Fig. 13G–H) has a single-wall thickness of 2.1 µm (Fig. 13I), while fragments of a larger, 3.5 mm, specimen have single-walls ca. 3.0 µm thick. The overall thickness of these specimens is therefore 4–6 µm, or, where folding has

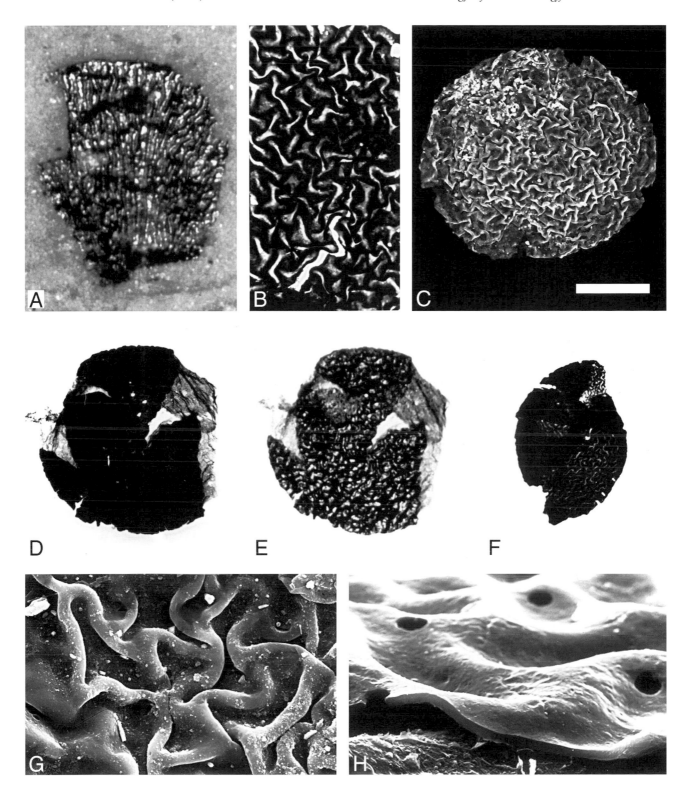

Fig. 12. Cerebrosphaera buickii n.gen., n.sp. From shales of the Algal Dolomite Member, Geerabukta (A, C, F–H), and the Lower Dolomite Member, Polarisbreen (B, D, E). Scale bar in C equals 200 μm for A, D–F; 80 μm for B, C; 15 μm for G; 7.5 μm for H. □A. HUPC 62761; 86-G-28-1BP; bedding-plane specimen with unusual sub-parallel/sub-radial wrinkling pattern. □B. HUPC 62762; P-2945-61M (N-30-2); transmitted light micrograph. □C. HUPC 62763; 86-G-33-2S; SEM. □D. HUPC 62713; P-2945-47M (S-33-4); holotype; transmitted light. □E. HUPC 62713 (same specimen as in D); reflected light. □F. HUPC 62764; 86-G-62-232M (R-15-4). □G. HUPC 62765; 86-G-33-1S; SEM; detail of surface convolutions. □H. HUPC 62766; 86-G-33-1S; SEM; detail of single-wall cross-section with rounded perforations.

taken place, up to 18 μm (these contacting layers are often fused and can thereby give an exaggerated measure of the original single-wall thickness).

All of the above measurements (of Svanbergfjellet material) were taken from well preserved specimens with relatively uneroded psilate surfaces. Most fossils, however, will have been subject to considerably greater degradation such that preserved wall thickness may be substantially reduced or lost altogether (e.g., Gussow 1973). In these cases vesicle-wall thickness and folding pattern can often be more reliably inferred from an examination of the enclosing sediments. During burial the 4–18 μm thick *Chuaria* interfered profoundly with the otherwise undisturbed bedding planes. The resulting imprints bear clear witness to the robust constitution of the interred vesicle, even in the absence of the original wall material (Fig. 8G–H; Walcott 1899; Ford & Breed 1973). Significant sedimentary molding is thus a reasonable criterion for assessing an assignment to *Chuaria*, at least in clastic sediments. *Leiosphaeridia wimanii* n.comb., for example, is comparable in overall dimensions to *Chuaria* but because of its much thinner walls does not impinge significantly upon its enclosing sediments (Fig. 13F; see below). The large (4.05 mm) spheroidal fossil in Fig. 8F is preserved solely as a thin organic film.

The walls of all unambiguous *C. circularis* in the Svanbergfjellet Formation are opaque to the transmitted light of a standard laboratory microscope. As discussed in the text (p. 12), this is not a direct function of wall thickness, nor can it be ascribed to differential taphonomy; rather, it appears to be an inherent property of *Chuaria* wall structure (Amard 1992). Opacity is also characteristic of the *Chuaria* type material (Walcott, 1899, p. 234) and conspecific *Fermoria* and *Protobolella* (Chapman 1935, p. 111), hence their early comparison with various phosphatic fossils (Rowell 1971; Ford & Breed 1973). No such confusion attended *Leiosphaeridia wimanii* n.comb. (= *C. wimani*) which is fundamentally translucent.

In the absence of any convincing data to the contrary, *Chuaria* is currently represented by a single species, *C. circularis*, although future analysis may yield additional well-defined groups. As with any taxonomic assignment, identification of a *Chuaria* species requires the positive documentation of diagnostic characters present in type materials; specimens lacking such features are best recorded as *Chuaria* sp. Thus, it is not always clear that fossils reported as *Chuaria circularis* are in fact resolvable to the species level (e.g., material reported by Hofmann, 1977), and it is indeed possible that the *Chuaria* of the oft-cited *Chuaria–Tawuia* assemblage (e.g., Duan 1982; Hofmann 1985a, b) may not, in fact, be *Chuaria circularis*. *Chuaria wimani* Brotzen, 1941, is clearly misplaced in the genus (see *Leiosphaeridia wimanii* n.comb.), and *C. globosa* Ogurtsova & Sergeev, 1989, is described as having a sheath, unlike the type species.

Chuaria circularis Walcott, 1899
Figs. 8G–H, 13G–I

Synonymy. – □1899 *Chuaria circularis*, nov. g. and sp. – Walcott, pp. 234–235, Pl. 27:12–13. □1935 *Fermoria minima*, gen. et sp. nov. – Chapman, pp. 115–116, Pl. 1:1, 3. □1935 *Fermoria granulosa*, gen. et sp. nov. – Chapman, p. 116, Pl. 1:2, 4; 2:5. □1935 *Fermoria capsella*, gen. et sp. nov. – Chapman, p. 117, Pl. 2:3, 4. □1935 *Protobolella jonesi*, gen. et sp. nov. – Chapman, pp. 117–118, Pl. 1:5, 6; 2:1. □1980 *Chuaria annularis* Zheng (sp. nov.) – Zheng, pp. 59–60, Pl. 1:3–4. □1984 *Chuaria minima* Chapman, 1935 emend. – Maithy & Shukla, pp. 146–148 (*partim*), Pl. 1:1–4; *non* Pl. 1:5–10. □1985 *Chuaria multirugosa* Du (sp. nov.) – Xing *et al.*, p. 70, Pl. 16:3.

Material. – Twenty (20) specimens: 19 on bedding planes, 1 isolated and mounted for SEM. From shale samples 86-G-28, 86-G-33, 86-G-61 and 86-G-62; An additional 65 specimens lack various features diagnostic of *C. circularis* and are here considered *Chuaria* spp.: 32 on bedding planes; 27 isolated and mounted on glass slides; 6 in bedding-parallel thin-section.

Diagnosis. – A species of *Chuaria* with well defined concentric wrinkles or folds around its periphery; 400–5000 μm in diameter.

Description. – Flattened spheroidal fossils preserved as thick-walled (2–3 μm single-wall thickness), opaque, carbonaceous vesicles, or as imprints/molds on shale bedding planes. Overall diameter 0.43–3.50 mm ($\bar{x} = 1.76$ mm; s.d. = 0.93 mm; $n = 20$). Concentric wrinkles/folds around the fossil periphery with sufficient relief to leave a substantial imprint on bedding planes. Surfaces psilate. Fractures, where present, tangentially rather than radially oriented. Envelope absent.

Discussion. – The circumscription of *C. circularis* has been variously discussed in terms of its overall diameter, wrinkling patterns, and wall thickness. Vidal & Ford (1985) further suggested that preservational mode, in this case as acid-resistant kerogen, be applied as a diagnostic character. While certainly useful in addressing the question of biogenicity, this latter condition is not applicable to the diagnosis of biologically meaningful fossil species (Sun 1987b, p. 351), and, in any event, the *original* carbonaceous constitution of *C. circularis* has already been diagnosed at the 'Group' (Acritarcha) level. In the Svanbergfjellet shales *C. circularis* occurs most commonly as bedding-plane imprints and compressions with the retention of various proportions of the carbonaceous wall (Fig. 8G–H), sometimes including an 'organic stain' (Fig. 8H). These structures are unquestionably biogenic and reliably assigned to *C. circularis*, but they are not acid resistant.

Concentric peripheral folding is certainly not unique to *C. circularis*, but it is a prominent feature of the type material

Fig 13. Trachyhystrichosphaera aimika. (A); *Leiosphaeridia* spp. (B–C); *Leiosphaeridia wimanii* (Brotzen) n.comb. (D–F); *Chuaria circularis* Walcott (G–I). From shales of the Algal Dolomite Member, Geerabukta. Scale bar in E equals 200 μm for A–C; 300 μm for D–H; 3 μm for I. □A. HUPC 62767; 86-G-62-230M (O-24-2); with perforations of the same size and arrangement as the processes of *Trachysystrichosphaera aimika.* □B. HUPC 62768; 86-G-62-172M (K-35-3); with distinctively perforated walls. □C. HUPC 62769; 86-G-62-71M (M-24-4); shagreenate surface texture. □D. HUPC 62770; 86-G-62-238M (K-20-2); with smooth-edged split. □E. HUPC 62771; 86-G-62-182M (S-27-0); with biologically mediated medial split. □F. HUPC 62772; 86-G-61-3BP; bedding-plane specimen with medial split. □G. HUPC 62773; 86-G-62-1S; SEM. □H. HUPC 62773 (same specimen in G); reflected light. □I. HUPC 62773; SEM; detail of G; cross-sectional fracture surface.

and offers an unambiguous means of subdividing the genus. Since the behavior of spheroids upon flattening broadly reflects the original nature of their walls (Harris 1974), specimens of (flattened) *Chuaria* lacking such folds might legitimately constitute a separate species (e.g., *Chuaria fermorei* Mathur, 1983); absence of folds suggests a considerably more plastic wall than in *C. circularis* (cf. Harris 1974: type 4 or 5 failure). In any event, *C. circularis* differs fundamentally from both *Cerebrosphaera buickii* n.sp., in which the thinner walls tend to fracture rather than fold, and *Leiosphaeridia wimanii* n.comb., where the thinner and translucent walls exhibit a pattern of folding no more regularly concentric than that found in smaller leiosphaerids.

Chuaria circularis is generally accepted as being less than 5 mm in diameter, but the lower size limit is not all clear. Ford & Breed (1973) have suggested an arbitrary lower size limit of 0.5 mm, and Jankauskas *et al.* (1987) one of 300 μm (subsequently altered to 1000 μm; Jankauskas *et al.* 1989), while Vidal & Ford (1985) have included specimens as small as 70 μm in the species. The Svanbergfjellet population of unambiguous *C. circularis* comprises 20 specimens ranging from 0.43 to 3.50 mm in diameter, with a mean of 1.81 mm. This compares well with the type material, and we suggest a lower size limit of 400 μm for the species. Thick-walled (or at least opaque) spheroids as small as 190 μm in diameter do occur in the Svanbergfjellet shales but they lack the diagnostic concentric wrinkling of *C. circularis*. If these forms can be shown to be *Chuaria* they await rigorous definition as a species other than *circularis*.

Genus *Comasphaeridium* Staplin, Jansonius & Pocock, 1965

Type species. – *Comasphaeridium cometes* (Valensi, 1949), p. 18.

Comasphaeridium sp.
Fig. 14E

Material. – Two (2) specimens from shale sample 86-G-62.

Description. – Thin-walled, spheroidal, carbonaceous vesicles covered with abundant, hair-like processes. Vesicle diameter ca. 50 μm; maximum overall diameter ca. 75 μm. Processes solid; straight to slightly curved; ca. 0.2 μm wide and up to 11 μm long.

Discussion. – *Comasphaeridium* is rare in the Svanbergfjellet assemblage, represented by just two specimens. These are broadly comparable to middle Silurian *C. sequestratus* Loeblich; however, final assignment awaits more and better preserved material. Svanbergfjellet *Comasphaeridium* are decidedly unlike the much larger (100–300 μm diameter) forms described from the Vendian of China (Zhang 1984; Yin L.

1985a) but closely resemble specimens from the Late Riphean Wynniatt Formation, Victoria Island, arctic Canada (Butterfield & Rainbird 1988).

Genus *Cymatiosphaeroides* Knoll, 1984, emend. Knoll, Swett & Mark, 1991

Type species. – *Cymatiosphaeroides kullingii* Knoll, 1984, p. 153.

Cymatiosphaeroides kullingii Knoll, 1984, emend.
Fig. 15A–E

Synonymy. – ☐1983 *Trachysphaeridium laufeldi* Vidal, 1976 – Knoll & Calder, p. 494, Pl. 60:1–2. ☐1984 *Cymatiosphaeroides kullingii* n.sp. – Knoll, p. 153, Fig. 9A–C. ☐1984 Unnamed form D – Knoll, p. 151, Fig. 9H. ☐1984 Pterospermopsimorphid form A – Knoll, p. 159, Fig. 7P–Q. ☐1989 *Cymatiosphaeroides kullingii* Knoll, 1984 – Allison & Awramik, p. 287, Fig. 11.1–11.3. ☐1991 *Cymatiosphaeroides kullingii* Knoll, 1984, emend. – Knoll *et al.*, p. 557, Fig. 4.4, 4.6.

Material. – Fifty seven (57) specimens: 56 in silicified carbonate; 1 in dolomite. From (partially) silicified carbonate samples 86-G-8, 86-G-9, 86-G-14, 86-G-15, P-2664 and P-2628.

Emended diagnosis. – A species of *Cymatiosphaeroides* with up to 12 enveloping membranes; vesicle diameter 30–350 μm; overall diameter 40–400 μm.

Description. – Spheroidal carbonaceous vesicles, 32–345 μm in diameter ($\bar{x} = 129$ μm; s.d. = 86.4 μm; $n = 51$), with abundant, narrow (ca. 1 μm) solid processes that support a single or multilaminated envelope. Maximum overall dimension 49–390 μm ($\bar{x} = 154$ μm; s.d. = 99 μm; $n = 53$). Up to 12 envelopes ($\bar{x} = 3$) surround a vesicle, sometimes producing a massive, space-filling mucilage in the interstices of intraclastic grainstone. Outer envelopes typically compressed and fused; occasionally separated from one another by process-supported intervals similar to that between the vesicle and inner envelope. Two vesicles rarely within a common envelope.

Discussion. – *Cymatiosphaeroides* and *C. kullingii* were recently rediagnosed by Knoll *et al.* (1991) who recognized the multiple layering of the envelopes. The Svanbergfjellet population generally supports these emendations, but reveals that both the number of enveloping layers (up to at least 12; Fig. 15A–B) and maximum dimensions (up to 400 μm) must be increased to encompass the considerable variability of the species.

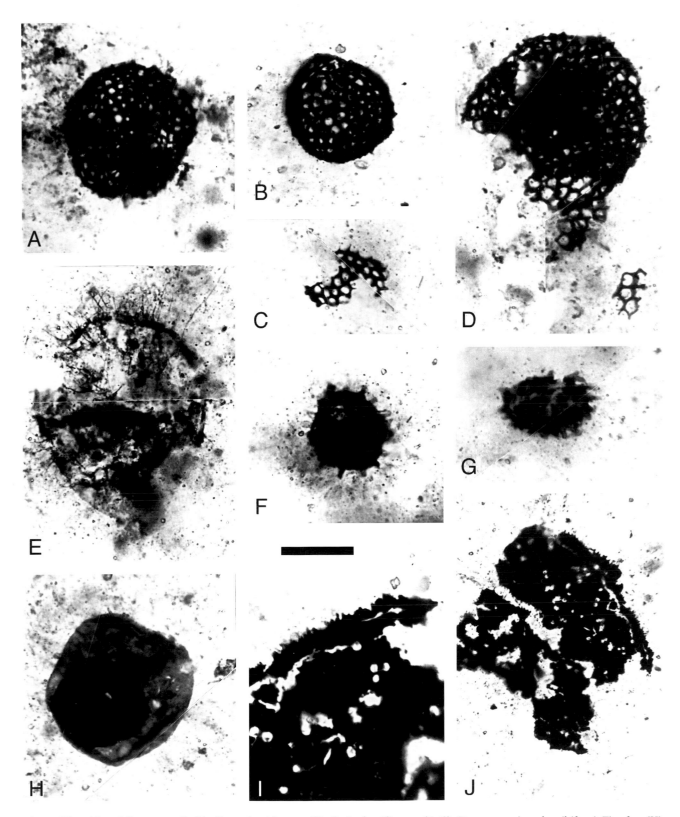

Fig. 14. Dictyotidium fullerene n.sp. (A–D); *Comasphaeridium* sp. (E); *Goniosphaeridium* sp. (F, G); *Pterospermopsimorpha pileiformis* Timofeev (H); *Gorgonisphaeridium* sp. (I, J). From shales of the Algal Dolomite Member, Geerabukta (A–H), and the Upper Limestone Member, Geerabukta (I, J). Scale bar in I equals 20 μm for A–I; 50 μm for J. □A. HUPC 62714; 86-G-62-45 (N-31-4); holotype; with membranes between the short protruding spines. □B. HUPC 62774; 86-G-62-51 (K-47-3). □C. HUPC 62775; 86-G-62-12 (T-27-3); isolated fragment. □D. HUPC 62776; 86-G-62-12 (N-32-1); broken specimen illustrating the characteristic, pre-burial brittleness. □E. HUPC 62777; 86-G-62-34 (Q-51-2). □F. HUPC 62778; 86-G-62-16 (S-16-4); with pylome(?) □G. HUPC 62779; 86-G-62-40 (R-51-4); with branched process(?). □H. HUPC 62780; 86-G-62-11 (V-36-1). □I. HUPC 62781; 86-G-63-3 (H-37-2); fragment. □J. HUPC 62781; detail of I.

Other than a slight mode at ca. 100 µm (maximum O.D), the relatively large population of *C. kullingii* in Svanbergfjellet cherts shows no obvious size-class separation. In combination with its often prolific envelope/mucilage production and wide range in superficial morphology, this continuous size distribution suggests that *C. kullingii* was an actively growing vegetative structure rather than a dormant cyst or spore. Such an interpretation is supported by evidence that *C. kullingii* actively attached itself to substrates (Fig. 15B; Knoll *et al.* 1991, Fig. 4.4) and the rare occurrence of two vesicles within a common outer envelope. *Cymatiosphaeroides kullingii* occurs primarily in shallow subtidal environments, often acting as a clast within microbialite grainstones; a single specimen embedded in dense filamentous mat (Fig. 15C) indicates its occasional transport to intertidal–supratidal settings (cf. Knoll *et al.* 1991).

In addition to the Svanbergfjellet occurrences, *C. kullingii* is found in the immediately overlying Draken Conglomerate Formation, the Hunnberg and Ryssö Formations on adjacent Nordaustlandet, and the upper Tindir Group of northwestern Canada. In silicified carbonate environments it always co-occurs with the large process-bearing acritarch *Trachyhystrichosphaera*. A variety of chemostratigraphic and biostratigraphic data point to this *Cymatiosphaeroides–Trachyhystrichosphaera* assemblage as diagnostic for the latter part of the Late Riphean (cf. Kaufman *et al.* 1992).

Genus *Dictyotidium*, Eisenack, 1955

Type species. – *Dictyotidium dictyotum* (Eisenack, 1938), p. 27–28.

Dictyotidium fullerene Butterfield, n.sp.
Fig. 14A–D

Holotype. – HUPC 62714, Fig. 14A; Slide 86-G-62-45, England-Finder coordinates N-31-4.

Type locality. – Algal Dolomite Member, Svanbergfjellet Formation, Geerabukta (79°35'30"N, 17°44'E); 55 m above base of member.

Material. – Nine (9) specimens from shale sample 86-G-62. Three (3) designated paratypes: HUPC 62774–62776, Fig. 14B–D.

Diagnosis. – A species of *Dictyotidium* with a robust reticulate framework defining 1.5–4 µm wide polygonal fields. Short solid processes project from the intersecting points of the polygons and support thin a membrane above the primary reticulum. Overall diameter, 30–60 µm.

Description. – Spheroidal carbonaceous microfossils, 30–60 µm in diameter ($\bar{x} = 33$ µm; s.d. $= 8$ µm; $n = 8$), defined by a robust reticulum of 1.5–4 µm wide polygonal fields; continuous vesicle wall not apparent. Short (ca. 1 µm) solid spines project from the intersecting points of the polygons and support a thin membrane above the primary reticulum. Reticulum often broken and occurring as isolated fragments.

Discussion. – *Dictyotidium fullerene* n.sp. is distinguished from other species in the genus by its robust reticulum and apparent absence of a vesicle wall within the polygonal fields. In combination, these two features probably contributed to its unusual pre-burial rigidity, as reflected in the occurrence of isolated fragments in thin section (Fig. 14C–D).

The short protruding spines of *D. fullerene* n.sp. are fully accommodated by *Dictyotidium*; however, the faint intervening membranes (Fig. 14A) are not stated as a feature of any previously named species. The Cambrian genus *Acrum* has broadly reminiscent membranes (Downie 1982, Fig. 3) but lacks spines. Despite this apparent conflict we consider the Svanbergfjellet fossils to be sufficiently close to other forms of *Dictyotidium* to be accommodated by the genus. An unnamed problematic microfossil from the Neoproterozoic Visingsö Beds of Sweden (Vidal 1976, Fig. 23A–D) has a brittle reticulate structure somewhat comparable to that in *D. fullerene*.

Etymology. – With reference to the reticulate structure of spheroidal molecular carbon – buckminsterfullerenes.

Genus *Germinosphaera* Mikhailova, 1986, emend.

Type species. – *Germinosphaera bispinosa* Mikhailova, 1986, p. 33.

Emended diagnosis. – Spheroidal vesicles with 1–6 open-ended, tubular and occasionally branched processes that communicate freely with the vesicle. Multiple processes usually restricted to a single 'equatorial' plane, but otherwise non-uniformly distributed on the vesicle.

Discussion. – *Germinosphaera* was established to accommodate distinctive, process-bearing spheroids from the Upper Riphean Dashkin Suite of Siberia; depending upon the number of processes, the four described specimens were placed in one of two species, *G. unispinosa* or *G. bispinosa*. The considerably larger Svanbergfjellet population (41 specimens) reveals that process number and length are highly variable in *Germinosphaera* and that this variation is almost certain to be intraspecific. Thus, the two very similarly sized original species are subsumed into one, *G. bispinosa*, while a Svanbergfjellet population with substantially larger vesicles represents a legitimate second species, *G. fibrilla* n.comb. A new, third form with a distinct wall texture and process habit may eventually warrant separate generic status but is here classified as *G. jankauskasii* n.sp.

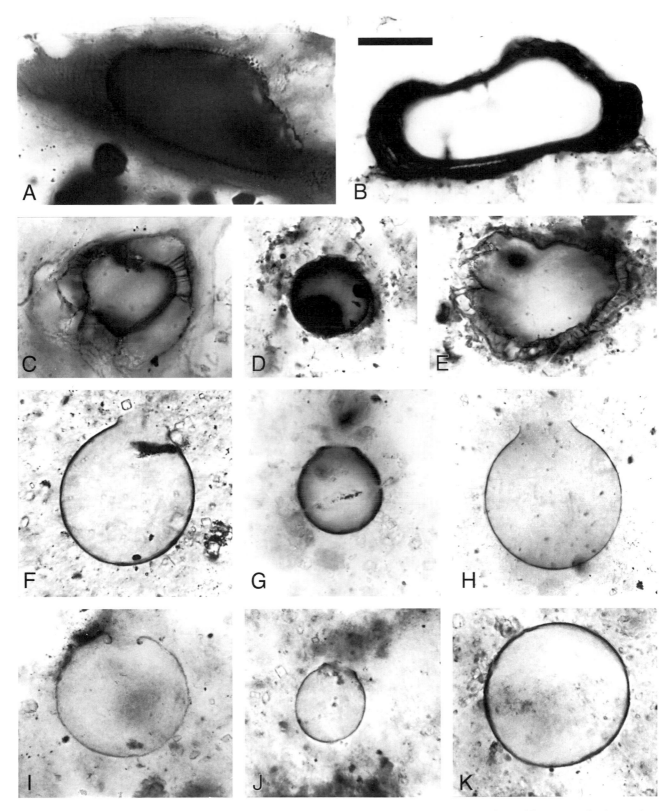

Fig. 15. Cymatiosphaeroides kullingii Knoll emend. (A–E) and *Osculosphaera hyalina* n.gen., n.sp. (F–K). From cherts of the Lower Dolomite Member, Geerabukta (A–E), and the Lower Limestone Member, Polarisbreen (F–K). Scale bar in B equals 25 μm for A, C; 75 μm for B; 60 μm for D–E; 50 μm for F–K. □A. HUPC 62782; 86-G-14-1A (J-43-1); with ca. 12 extracellular layers; in intraclastic grainstone. □B. HUPC 62783; 86-G-14-1A (F-72-4); with ca. 12 extracellular layers; attached to intraclast. □C. HUPC 62784; 86-G-15-1 (L-64-3); within a microbialite intraclast. □D. HUPC 62785; 86-G-8-2B (U-64-0); within a laminated intraclast. □E. HUPC 62786; 86-G-8-2B (U-61-4); within a laminated intraclast. □F. HUPC 62716; P-3085-1A (P-43-2); holotype. □G. HUPC 62787; 86-P-82-1B (X-47-2). □H. HUPC 62788; P-3085-1E (U-48-4). □I. HUPC 62789; P-3085-1C (G-63-3); with 'oral' collar curled inward. □J. HUPC 62790; P-3085-1D (P-65-4). □K. HUPC 62791; P-3085-1E (E-62-2); co-occurring spheroid of the same size and structure as *O. hyalina*.

The more or less random disposition of hollow, open-ended processes characteristic of *Germinosphaera* suggests it was an actively growing vegetative structure, rather than a dormant cyst or spore (which are typically self-contained and regularly symmetrical). These fossils have a close morphological (though not necessarily taxonomic) analogue in the germinating zoospores of the modern xanthophyte alga, *Vaucheria*. The large (ca. 100 μm diameter) asexual spores of *Vaucheria* establish new filamentous thalli by germinating one to several non-septate and sometimes branching primordia (Fig. 16G). Under this 'germinating zoospore' interpretation *Germinosphaera* would derive from relatively undistinguished spheroidal acritarchs, and the typical open-endedness of its processes from simple mechanical breakage.

Germinosphaera bispinosa Mikhailova, 1986, emend.

Fig. 16D–E

Synonymy. – □1986 *Germinosphaera bispinosa* Mikhailova sp. nov. – Mikhailova, p. 33, Fig. 6. □1986 *Germinosphaera unispinosa* Mikhailova sp. nov. – Mikhailova, p. 33, Fig. 5.

Holotype. – No 882/2; Krasnoyarsk region, R. Uderei; Upper Rifean, Dashkin Suite; Mikhailova 1986, Fig. 6.

Material. – Fourteen (14) specimens from shale sample 86-G-62.

Emended diagnosis. – A species of *Germinosphaera* with psilate vesicles 13–35 μm in diameter. Processes 2.5–3.5 μm wide and, when multiple, arranged equatorially on the vesicle.

Description. – Thin-walled, psilate and spheroidal carbonaceous vesicles, 18–35 μm in diameter ($\bar{x}$ = 25 μm; s.d. = 5 μm; n = 14), with 1–4 tubular, filamentous processes, 2.5–3.5 μm wide and up to 185 μm long. Process cavities unobstructed, both proximally with the vesicle and terminally to the outside. Multiple processes arranged equatorially on a vesicle.

Discussion. – A majority of Svanbergfjellet *G. bispinosa* n.sp. bear a single filamentous process; however, one otherwise indistinguishable specimen has four such extensions arranged at ca. 90° intervals around the flattened vesicle perimeter. Specimens of *G. bispinosa* with multiple processes also occur in the broadly correlative Wynniatt Formation, arctic Canada (Butterfield & Rainbird, unpublished data).

Germinosphaera fibrilla (Ouyang, Yin & Li, 1974) Butterfield, n.comb.

Fig. 17A–H

Synonymy. – □1974 *Ooidium fibrillum* (sp. nov.) – Ouyang *et al.*, p. 120, Pl. 47:4. □1974 *Archaeohystrichosphaeridium*

truncatum (sp. nov.) – Ouyang *et al.*, p. 77, Pl. 27:19. □1985 *Phycomycetes* sp. – Yin C., Pl. 4:19. □1989 *Germinosphaera tadasii* A. Weiss, sp. nov. – Jankauskas *et al.*, p. 143, Pl. 47:3–4 (*partim*); *non* Pl. 47:5.

Holotype. – Ouyang, Yin & Li 1974, Pl. 47:4.

Material. – Eighteen (18) specimens from shale sample 86-G-62.

Diagnosis. – A species of *Germinosphaera* with psilate vesicles 45–125 μm in diameter. Processes 3–5 μm wide and, when multiple, arranged equatorially on the vesicle.

Description. – Thin-walled, psilate and spheroidal carbonaceous vesicles, 49–120 μm in diameter ($\bar{x}$ = 87 μm; s.d. = 21 μm; n = 18), with 1–4 filamentous, tubular and occasionally branched processes, 3–5 μm wide and up to 90 μm long. Process cavities unobstructed, both proximally with the vesicle, and terminally to the outside. Multiple processes arranged equatorially on a vesicle, although not always evenly-spaced.

Discussion. – Unlike the type species, *G. fibrilla* n.comb. commonly bear 2–4 processes. In specimens with more than two processes, their diagnostic equatorial positioning also becomes clear, with the third (and fourth, if present) process never originating from within the perimeter of the flattened vesicle. Such an orientation suggests that *G. fibrilla* was a benthic organism in which the primary growth plane was horizontal; a similar arrangement is observed in the germinating zoospores of surface-growing *Vaucheria*. Branching processes in *G. fibrilla* (Fig. 17F) lend further support to the 'germinating zoospore' interpretation (cf. modern *Vaucheria*; Fig. 16G).

The type specimen of *G. fibrilla* n.comb., from the Sinian of southwestern China, was clearly misplaced as a species of *Ooidium*. With its ca. 85 μm diameter vesicle and three, ca. 4 μm wide, equatorially arranged processes it is all but indistinguishable from the Svanbergfjellet material.

Germinosphaera jankauskasii Butterfield, n.sp.

Fig. 16A–C

Holotype. – HUPC 62715, Fig. 16A; Slide 86-G-62-12M, England-Finder coordinates L-31-3.

Type locality. – Algal Dolomite Member, Svanbergfjellet Formation, Geerabukta (79°35'30"N, 17°44'E); 55 m above base of member.

Material. – Six (6) specimens from shale sample 86-G-62. Two (2) designated paratypes: HUPC 62792–62793, Fig. 16B–C.

Diagnosis. – A species of *Germinosphaera* with shagrinate walls and vesicles 45–90 μm in diameter. Processes 5–10 μm wide and randomly distributed.

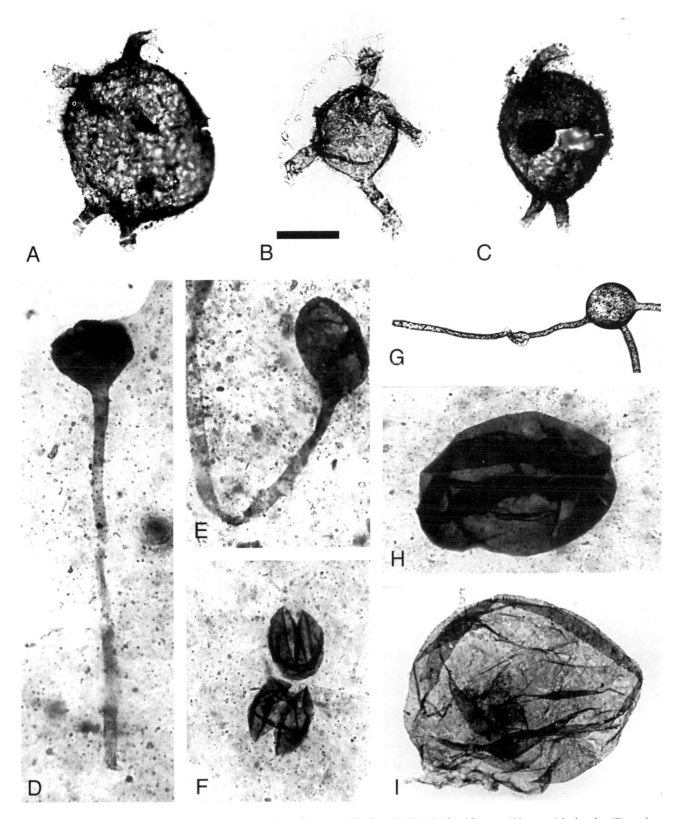

Fig. 16. *Germinosphaera jankauskasii* n.sp. (A–C); *Germinosphaera bispinosa* Mikhailova (D, E); *Leiosphaeridia crassa* (Naumova) Jankauskas (F); modern *Vaucheria sessilis*, germinating zoospore showing multiple and occasionally branched filament primordia (G); *Leiosphaeridia jacutica* (Timofeev) Mikhailova & Jankauskas (H); and *Leiosphaeridia tenuissima* Eisenack (I). All fossils from shales of the Algal Dolomite Member, Geerabukta. Scale bar in B equals 30 μm for A–C; 20 μm for D–F; 140 μm for G; 40 μm for H–I. □A. HUPC 62715; 86-G-62-12M (L-31-3); holotype. □B. HUPC 62792; 86-G-62-30M (J-33-1). □C. HUPC 62793; 86-G-62-155M (N-35-1). □D. HUPC 62794; 86-G-62-28 (O-19-2). □E. HUPC 62795; 86-G-62-14 (J-41-2). □F. HUPC 62796; 86-G-62-6 (E-25-4); with medial splits. □G. *Vaucheria sessilis.* □H. HUPC 62797; 86-G-62-10 (S-33-3). □I. HUPC 62798; 86-G-62-71M (M-16-3).

Description. – Shagrinate, spheroidal, carbonaceous vesicles, 46–86 μm in diameter ($\bar{x}$ = 71 μm; s.d. = 14 μm; n = 6), with 2–7 slightly capitate, hollow processes, 5–10 μm wide and up to 32 μm long. Process cavities unobstructed, both proximally with the vesicle and distally to the outside. Processes randomly positioned on a vesicle; not limited to a single plane.

Discussion. – *Germinosphaera jankauskasii* n.sp. is distinguished from other species of *Germinosphaera* by its shagrinate wall texture (possibly the remnants of an outer mucilaginous layer), significantly broader processes, and a random positioning of processes on the vesicle (*vs.* constrained to one plane). This latter condition is best observed in the type specimen (Fig. 16A) where, in addition to the four 'equatorial' processes, there are three that clearly originate from within the perimeter of the flattened vesicle.

The indeterminate number and positioning of processes in *G. jankauskasii* n.sp. suggest a functional interpretation similar to that of other *Germinosphaera* species, i.e. a germinating propagule broadly comparable to the zoospores of *Vaucheria* (Fig. 16G). Unlike those of the other forms, however, *G. jankauskasii* primordia appear to have had no preferred orientation, and growth evidently proceeded in three dimensions.

Etymology. – In recognition of Lithuanian paleontologist Dr. Tadas Jankauskas.

Genus *Goniosphaeridium* Eisenack, 1969

Type species. – *Goniosphaeridium polygonale* (Eisenack, 1931), p. 113.

Goniosphaeridium sp.
Fig. 14F–G

Material. – Two (2) specimens from shale sample 86-G-62.

Description. – Spheroidal carbonaceous vesicles, ca. 24 μm in diameter, with numerous, evenly spaced, evexate (blunt-tipped) processes, ca. 1.5 μm broad by 3 μm long. Processes apparently hollow with cavities that communicate freely with the vesicle lumen. One specimen with a pylome-like opening 5 μm in diameter; another with a branched(?) process.

Discussion. – These small, 'Paleozoic-aspect' acanthomorphic acritarchs (p. 17) do not readily recall any named species of *Goniosphaeridium*; they do, however, conform to its generic diagnosis of psilate or shagrinate vesicles with hollow, unconstricted, and simple processes. Closer overall comparisons can be made with species of *Baltisphaeridium* (e.g., *B. brevispinosum* (Eisenack)) or *Gorgonisphaeridium* (e.g., *G. echinodermum* (Stockmans & Willière)), but the processes of these forms are, respectively, isolated from the central vesicle, and solid. A single, apparently branching process in one Svanbergfjellet specimen (Fig. 14G) and a pylome-like structure in the other (Fig. 14F) further complicate the taxonomy of these fossils.

Genus *Gorgonisphaeridium* Staplin, Jansonius & Pocock, 1965

Type species. – *Gorgonisphaeridium winslowii* Staplin, Jansonius & Pocock, 1965, p. 193.

Gorgonisphaeridium sp.
Fig. 14I–J

Material. – One (1) specimen in shale sample 86-G-63.

Description. – Approximately 1 quadrant of a thick-walled, spheroidal, carbonaceous vesicle with an echinate surface. Processes solid, tapering from 1.5 to 1.0 μm and ca. 2 μm long. Full vesicle diameter calculated to be ca. 250 μm.

Discussion. – This isolated fragment compares with the single '*Gorgonisphaeridium maximum*' described from the overlying Draken Conglomerate (Knoll *et al.* 1991). The principal difference lies in the shorter processes of the Svanbergfjellet specimen, possibly a consequence of transport erosion. Poor preservation precludes any definitive taxonomic assignment.

Genus *Leiosphaeridia* Eisenack, 1958

Type species. – *Leiosphaeridia baltica* Eisenack, 1958, p. 2–3.

Discussion. – Jankauskas *et al.* (1987, 1989) offered a comprehensive revision of Proterozoic leiosphaerid acritarchs, dividing the smooth-walled forms into four basic species based on wall thickness and size class: *L. minutissima* – thin-walled, <70 μm; *L. tenuissima* – thin-walled, 70–200 μm; *L. crassa* – thicker-walled, < 70 μm; *L. jacutica* – thicker-walled, 70–800 μm. There are some difficulties with this taxonomy, particularly in dealing with larger forms, and in assessing wall thicknesses; overall, however, it is proving to be both a useful and not altogether artificial classification. All four basic species are found in the Svanbergfjellet shales, and there does appear to be a break in size-frequency distributions on or around 70 μm (Fig. 4).

Leiosphaeridia crassa (Naumova, 1949) Jankauskas, 1989
Figs. 16F, 23K

Material. – Abundant in most fossiliferous shale samples; 1220 measured specimens.

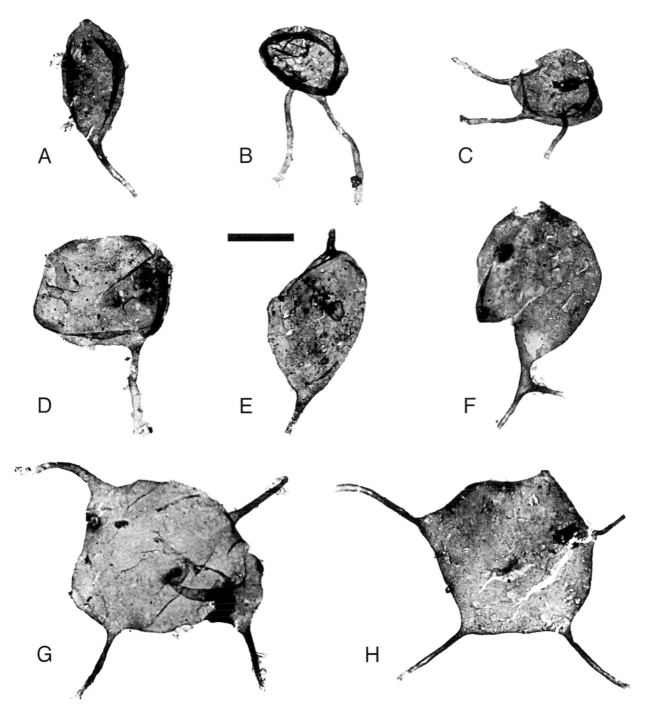

Fig. 17. Germinosphaera fibrilla (Ouyang, Yin & Li) n.comb. From shales of the Algal Dolomite Member, Geerabukta. Scale bar in 5 equals 50 μm. □A. HUPC 62799; 86-G-62-97M (L-38-3). □B. HUPC 62800; 86-G-62-24M (H-31-2). □C. HUPC 62801; 86-G-62-27M (O-38-4). □D. HUPC 62802; 86-G-62-23M (J-30-3). □E. HUPC 62803; 86-G-62-25M (N-28-3). □F. HUPC 62804; 86-G-62-100M (O-29-3); with branching process. □G. HUPC 62805; 86-G-62-28M (O-30-0). □H. HUPC 62806; 86-G-62-33M (K-36-3); holotype.

Discussion. – Leiosphaerid acritarchs less than 70 μm in diameter are abundant in Svanbergfjellet shales. In the Algal Dolomite Member (sample 86-G-62) they show a clearly defined, right-skewed size-frequency distribution with a single mode at ca. 17 μm ($\bar{x} = 19.1$ μm; s.d. = 10.7; $n = 1065$; Fig. 4). Leiosphaerid size-frequency in the Lower Dolomite

Member shale (sample P-2945) is distinct but likewise shows a break at ca. 70 μm (Fig. 4).

Leiosphaeridia crassa occurs variously as isolated individuals, large populations covering entire bedding planes, or very localized associations; in this latter conformation it clearly grades into sheet-like *Ostiana.* Some specimens show release structures (Fig. 16F), suggesting their role as cysts or, alternatively, as vegetative cells mother cells similar to those found in modern *Chlorella* (cf. Butterfield & Chandler 1992). Others were presumably the prerequisite stage to small germinosphaerids (Figs. 16D–E, 17).

Leiosphaeridia jacutica (Timofeev, 1966) Mikhailova & Jankauskas, 1989

Fig. 16H

Material. – Two hundred fifty two (252) measured specimens: 150 isolated by acid maceration; 102 in bedding-parallel thin section. From shale samples 86-G-62, 86-G-61, 86-G-33, 86-G-28 and P-2945.

Discussion. – Svanbergfjellet *L. jacutica* range from 71 to 796 µm in diameter and a population isolated by acid maceration shows a broad, right-skewed size distribution centered on ca. 150 µm ($\bar{x}$ = 232; s.d. = 116; n = 150). While this broad modality suggests the prevalence of a single natural taxon, *L. jacutica* is likely to encompass a disparate variety of fossil organisms. In the Svanbergfjellet assemblage such a form would clearly have given rise to *Germinosphaera fibrilla* n.comb., *Osculosphaera hyalina* n.sp. (Fig. 15K), and *Trachyhystrichosphaera aimika* (Fig. 18A); *T. aimika* may also 're-turn' to a leiosphaerid habit through erosion of its processes (e.g., Fig. 13A).

Leiosphaeridia tenuissima Eisenack, 1958

Fig. 16I

Material. – Twenty one (21) specimens from shale samples 86-G-62, 86-G-63 and P-2945.

Discussion. – These thin-walled leiosphaerids are distinguished from *L. jacutica* by the relative transparency and (original) flexibility of their walls. Svanbergfjellet *L. tenuissima* range from 92 to 624 in diameter (considerably larger than the 200 µm upper limit suggested in Jankauskas *et al.* 1989) and show a right-skewed size-frequency distribution centered broadly on 150 µm ($\bar{x}$ = 237 µm; s.d. = 129; n = 21). The marked similarity between the size distributions of *L. tenuissima* and *L. jacutica* casts some doubt on the purported distinction between the two species.

Leiosphaeridia wimanii (Brotzen, 1941) Butterfield, n.comb.

Fig. 13D–F

Synonymy. – □1894 Das Fossil aus der Visingsögruppe – Wiman, Pl. 5:1–5. □1941 *Chuaria wimani* n.nom. – Brotzen, pp. 258–259. □1969 *Kildinella magna* sp.n. – Timofeev, p. 14, Pl. 6:4–5. □1991 *Shouhsienia shouhsienensis* Xing

(Hsing) – Zhang *et al.*, p. 120, Pl. 1:16–26 (see for extended synonymy).

Typification and orthography. – Brotzen (1941) provided a new name for 'Das Fossil aus der Visingsögruppe' illustrated by Wiman (1894) but failed to specify a type specimen. We here designate the lectotype (ICBN Article 7.4) to be: Uppsala University, Visingsö Collection. 1-9, (Wiman 1894, Pl. 5:3; Ford & Breed 1973, Pl. 62:3); reposited in the Palaeontological Museum, Uppsala, PMU Ög 12. The orthography of the specific epithet, *wimani*, is corrected in accordance with ICBN Articles 32.5 and 73.10.

Material. – Fourteen (14) specimens from shale samples 86-G-62, 86-G-61, 86-G-33 and P-2945.

Diagnosis. – A species of *Leiosphaeridia* 800–2500 µm in diameter. Walls translucent, psilate to finely textured, commonly with medial-split-release structures.

Description. – Psilate or finely textured, spheroidal, carbonaceous vesicles, 0.76–2.48 mm diameter ($\bar{x}$ = 1.27 mm; s.d. = 0.44 mm; n = 14), commonly with a single medial split. Walls thin, leaving little or no imprint on enclosing sediments, and typically translucent in transmitted light. Folds, when present, in various orientations.

Discussion. – The stated size range for *L. wimanii* n.comb. is admittedly arbitrary; however, these large, relatively thin-walled spheroids are clearly something other than the tail end of the *L. jacutica* distribution. The vesicles, for example, commonly show biologically mediated splits, a feature illustrated by the smooth, rounded edges of the opening (Fig. 13E–13F; Wiman, 1894; Timofeev 1969; Zhang *et al.* 1991) and evidence that compressional folding took place after the wall had ruptured (Fig. 13D).

The type material of *L. wimanii* n.comb., from the Late Proterozoic Visingsö Beds (Wiman, 1894), was initially described as a species of *Chuaria* (Brotzen 1941). Eisenack (1951) subsequently referred it to *Leiosphaera* (= *Leiosphaeridia*), although he later reversed this decision, considering it to be a chitinous foram (Eisenack 1966). More recently, Ford & Breed (1973), Vidal & Ford (1985), and Jankauskas *et al.* (1989) subsumed *C. wimani* (including *Kildinella magna* Timofeev 1969) into *C. circularis*. These authors notwithstanding, a close analysis of the essential features of *Chuaria circularis* (see above) and those of the Visingsö fossils originally described as *C. wimani* shows the two forms to have little in common other than their millimetric dimensions. *Chuaria* is always opaque and is sufficiently thick-walled to leave a substantial imprint on bedding planes; excystment/release structures do not occur in the type material (Ford & Breed 1973, p. 546) or in the Indian (Chapman 1935) or Svanbergfjellet populations. By contrast, the Visingsö fossils often exhibit medial splits, are translucent through three or more superimposed layers (Wiman 1894, Pl. 5:1–3; Eisenack 1966, Fig. 2; Timofeev 1970, Pl. 1A, B), and leave only the slightest, if any, perceptible imprint on bedding-plane surfaces; SEM investigation of a split 1.25-mm-diameter Visingsö '*C. wimani*' (sample courtesy of G. Vidal) showed its wall to be only 1 µm thick, half that of a considerably smaller Svanbergfjellet *C. circularis* (Fig. 13G–I). Apart from their greater overall diameter, these fossils are indistinguishable from other leiosphaerids and are therefore revised as *L. wimanii* n.comb. *Leiosphaeridia wimanii* from the Svanbergfjellet Formation share all the features of the Visingsö type material.

Leiosphaeridia spp.

Fig. 13B–C

Discussion. – The Svanbergfjellet assemblage undoubtedly includes many more leiosphaerid species than can be determined from size and wall opacity/thickness alone. The difficulty lies in diagnosing their essential features. Two leiosphaerid forms are readily distinguished in the Algal Dolomite Member shales, but only on the basis of what appear to be taphonomically induced surface textures. In the one case, dark-walled vesicles, 181–488 µm in diameter ($\bar{x}$ = 319 µm; s.d. = 116 µm, n = 6), are perforated by numerous holes with a distinctive bridged habit (Fig. 13B); in the other, thin-walled vesicles, 344–1325 µm in diameter ($\bar{x}$ = 798 µm; s.d. = 333 µm, n = 7), have a shagrinate to velutinous wall structure and commonly include a central, longitudinal crease or inclusion (Fig. 13C). These taphonomic(?) fabrics may well represent underlying taxonomic differences but, in the absence of supporting primary features, cannot be legitimately applied to taxonomic diagnoses.

Genus *Osculosphaera* Butterfield, n.gen.

Type species. – *Osculosphaera hyalina* n.sp.

Diagnosis. – Psilate, hyaline spheroidal vesicles with a single, rimmed, circular opening, 25–50% the diameter of the vesicle.

Discussion. – While *Osculosphaera* n.gen. clearly derives from a simple spheroidal precursor, its prominent rimmed opening represents a substantial qualitative, and therefore generic-level distinction from *Leiosphaeridia*. Comparable acritarchs have been assigned to the genus *Palaeostomocystis* (cf. *P. apiculata* Cookson & Eisenack, 1960; *P. laevigata* Drugg, 1967; *P. angoformis* Yin, 1985a); however, the type species, *P. reticulata* Deflandre, 1937, is heavily ornamented and quite unlike these other taxa. *Osculosphaera* n.gen. is established to accommodate a distinctive population of porate Svanbergfjellet acritarchs, along with a closely comparable but smaller form from Late Riphean shales of the Southern Urals – *Leiosphaeridia kulgunica* Jankauskas, 1980(a) – and, perhaps, the above-mentioned species of *Palaeostomocystis*. The proposed *Osculosphaera kulgunica*

(Jankauskas, 1980) n.comb. shares with *O. hyalina* n.sp. psilate spheroidal vesicles, a well defined circular opening of the same relative proportions, and somewhat rigid walls. This latter character is expressed in flattened *O. kulgunica* n.comb. compressions by their characteristic radial fractures, and in the three-dimensionally preserved *O. hyalina* n.sp. by its consistent sphericity and rare angular indentations.

Svanbergfjellet *Osculosphaera* n.gen. are more or less flask-shaped in lateral aspect (Fig. 15F–H, J), inviting comparison with vase-shaped microfossils such as *Melanocyrillium* Bloeser, 1985. This latter form, however, is characterized by very thick walls and various circumoral ornamentation. *Hyalocyrillium* Allison, 1989, more closely approximates the wall structure of *Osculosphaera*, but its long, more or less cylindrical habit is entirely distinct from spheroidal *Osculosphaera*.

Etymology. – From the Latin *osculum*, mouth, and *sphaera*, ball.

Osculosphaera hyalina Butterfield, n.sp.

Fig. 15F–J

Holotype. – HUPC 62716, Fig. 15F; Slide P-3085-1A, England-Finder coordinates P-43-2.

Type locality. – Lower Limestone Member, Svanbergfjellet Formation, Polarisbreen (79°10'N, 18°12'E); 26 m above the base of the *Minjaria* biostrome.

Material. – Eleven (11) specimens in chert samples P-3085 and 86-P-82 (from a single horizon). Seventy (70) associated entire spheroids. Four (4) designated paratypes: HUPC 62787–62790, Fig. 15G–J.

Diagnosis. – A species of *Osculosphaera* with hyaline walls; vesicles 35–150 µm in diameter.

Description. – Psilate, hyaline, spheroidal vesicles, 35–131 µm in diameter ($\bar{x}$ = 78 µm; s.d. = 29 µm; n = 11), with a single prominent circular opening, 30–45% the diameter of the vesicle. Wall around the opening typically projected outward forming a short 'oral collar'; rarely curled inward. A single specimen with three spheroids, ca. 13 µm in diameter, within the vesicle.

Discussion. – *Osculosphaera hyalina* n.sp. co-occurs with a population of undifferentiated spheroids (Fig. 15K), closely comparable in wall structure and size-frequency distribution (22–163 µm diameter; $\bar{x}$ = 79 µm; s.d. = 35 µm; n = 70). These are clearly the antecedent form of *O. hyalina*, although in a strict form taxonomy they would presumably be assigned to species of *Leiosphaeridia*; it is the examination *in situ* (i.e. in thin-section) of the Svanbergfjellet population that allows this finer resolution. For the purposes of biostratigraphy (the principal application of acritarch taxonomy) such biologically interesting details are not of direct relevance, and, since any number of forms can derive from simple spheroids (p. 17), this ontogenetic variant is not included in the diagnosis. The presence of multiple small spheroids within one *O. hyalina* vesicle points to its role as a reproductive body, with the pore serving as a release structure.

There is insufficient material in the 'oral collar' of *O. hyalina* n.sp. to recover the antecedent spheroid in its entirety, an indication that some of the wall has been subject to biologically mediated removal. Actual restructuring of the wall, however, appears to have been minimal, with the 'collar' typically showing no secondary reinforcement. The principal alteration may have been an induced flexibility in the area immediately around the opening (Fig. 15I).

Etymology. – From the Greek *hyalinos*, of glass, in reference to the translucent and somewhat rigid walls.

Genus *Pterospermopsimorpha* Timofeev, 1966

Type species. – *Pterospermopsimorpha pileiformis* Timofeev, 1966, p. 34.

Discussion. – Probable eukaryotic spheroids with envelopes are relatively uncommon in Svanbergfjellet biota, but given their wide range in size (17–213 μm) and wall structure they are sure to represent a number of *Pterospermopsimorpha* species. Conversely, the marked size discrepancy between the inner and outer vesicles illustrates the potential for an artificial inflation of leiosphaerid species diversity. Some silicified pterospermopsimorphs are also likely to be degradational (or developmental) variants of *Cymatiosphaeroides* (see above).

Pterospermopsimorpha pileiformis Timofeev, 1966
Fig. 14H

Material. – One (1) specimen in shale sample 86-G-62. Five (5) additional specimens assigned to *Pterospermopsimorpha* spp. from shale samples 86-G-62, 86-G-61 and P-2945.

Description. – Smooth-walled spheroid preserved within a larger smooth-walled spheroidal envelope. Inner vesicle 25 μm in diameter; outer vesicle 43 μm in diameter; both with simple folds.

Genus *Trachyhystrichosphaera* Timofeev & Hermann, 1976, emend.

Synonymy. – ☐1976 *Trachyhystrichosphaera* Timofeev et Hermann gen. n. – Timofeev *et al.*, p. 48. ☐1976 *Nucello-*

hystrichosphaera Timofeev et Hermann gen. et sp. n. – Timofeev *et al.*, p. 47. ☐*non* 1980 *Trachyhystrichosphaera bothnica* n.sp. – Tynni & Donner, p. 11, Pl. 2:15–17. ☐*nec* 1989 *Trachyhystrichosphaera parva* Mikhailova, sp. nov. – Jankauskas *et al.*, p. 47, Pl. 2:2–3. ☐*nec* 1989 *Trachyhystrichosphaera truncata* Hermann et Jankauskas, sp. nov. – Jankauskas *et al.*, p. 48, Pl. 2:5–6. ☐*nec* 1992 *Trachyhystrichosphaera? anelpa* sp. nov. – Zang & Walter 1992a, pp. 110–112, Fig. 82A–E. ☐*nec* 1992 *Trachyhystrichosphaera hystricosa* sp. nov. – Zang & Walter 1992a, p. 112, Fig. 84A–H. ☐*nec* 1992 *Trachyhystrichosphaera triangula* sp. nov. – Zang & Walter 1992a, p. 112, Fig. 86A–H. ☐*nec* 1992 *Trachyhystrichosphaera tribulosa* sp. nov. – Zang & Walter 1992a, pp. 112–114, Fig. 87A–H.

Type species. – *Trachyhystrichosphaera aimika* Hermann, 1976, p. 48 (*in* Timofeev *et al.* 1976).

Emended diagnosis. – Relatively large spheroidal vesicles with one to many irregularly distributed hollow processes, 3–8 μm in diameter. Vesicle wall psilate to (degradationally) shagrinate, or echinate. Processes communicate freely with the vesicle cavity but otherwise highly heteromorphic: tubular, conical, capitate, bifurcate, and various other forms. Outer sheath/envelope variable: absent to diaphanous to robust; occasionally bi-layered in silicified specimens.

Discussion. – *Trachyhystrichosphaera* is distinguished from other acanthomorphic acritarchs by its marked irregularity. The processes are highly heteromorphic (within a limited size range), vary widely in number, and show no regularity in their distribution on a vesicle; similarly, an outer envelope can take on a variety of habits or be absent entirely (see Vidal *et al.* 1993, Pl. 1:6). Recognition of systematic trends within and between populations (including a size range of well over an order of magnitude; Knoll *et al.* 1991) nonetheless suggests a representation of relatively few natural taxa exhibiting high intraspecific variability. The source of such variation might be ontogenetic, environmental, or, most probably, some combination of the two. Given the presence of vesicles with single and/or very small processes (Fig. 18A, K), it would appear that *Trachyhystrichosphaera* individuals began their life cycle as simple leiosphaerids which subsequently, and sequentially, differentiated their various processes and envelopes. Envelope or sheath formation appears to have followed the appearance of at least some of the processes, as evinced by the absence of envelopes in those specimens with few or predominately short processes (Figs. 18A–C, G, 19A, B) and the common protrusion of processes through a relatively proximal sheath (Fig. 18D, I, J). Those vesicles with the greatest number of processes are typically those with the most uniform and uniformly arranged processes, and they also exhibit the most fully developed envelopes (Figs. 18D, 19D, E, H; 19D has subsequently lost its sheath); this is conceivably the most 'mature' stage. Even so, essentially indistinguishable forms can be found with or without envelopes (Fig. 19C, E; Pjatiletov 1988).

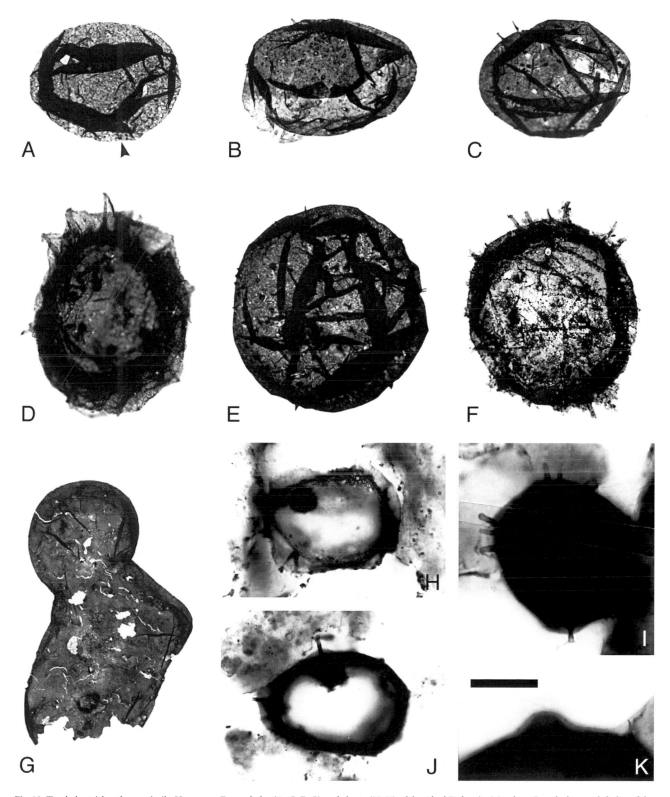

Fig. 18. Trachyhystrichosphaera aimika Hermann. From shales (A–C, E, G) and cherts (H–K) of the Algal Dolomite Member, Geerabukta, and shales of the Lower Dolomite Member, Polarisbreen (D, F). Scale bar in K equals 75 µm for A–F, H–J; 150 µm for G; 25 µm for K. □A. HUPC 62807; 86-G-62-163M (B-31-2); with small conical process at the arrow. □B. HUPC 62808; 86-G-62-159M (O-15-2). □C. HUPC 62809; 86-G-62-32M (F-45-4). □D. HUPC 62810; P-2945-1M (K-30-3); with partial envelope. □E. HUPC 62811; 86-G-62-31M (G-33-0); with heterotrophic processes. □F. HUPC 62812; P-2945-2M (L-33-1). □G. HUPC 62813; 86-G-62-158M (L-26-3); with medial constriction and inconspicuous processes. □H. HUPC 62814; 86-G-15-2A (S-35-1); with bifurcate processes. □I. HUPC 62815; 86-G-15-2A (O-38-3); attached to intraclast. □J. HUPC 62816; 86-G-15-2A (Q-35-2). □K. HUPC 62816; detail of J; with incipient process and three discrete wall layers.

This marked variation of *Trachyhystrichosphaera* points clearly to its active and environmentally interactive physiology; it was not a dormant body. Broadly comparable structures are found as the degradationally resistant phycomata of modern prasinophyte green algae, which are likewise capable of substantial increases in size (Tappan 1980; Knoll *et al.* 1991). Such features, however, provide little in the way of taxonomic criteria and, in the absence of more detailed comparisons, the relationships of *Trachyhystrichosphaera* remain unresolved.

Differential taphonomy adds a further level of 'variation' to *Trachyhystrichosphaera*. Thus, many of the supposed distinctions in wall texture can be attributed to the variable preservation of an extracellular sheath (Pjatiletov 1988; Jankauskas *et al.* 1989, Pl. 1:7, 8). Moreover, the unusual terminal expansion of some processes appears to coincide with their penetration through and beyond the envelope (Jankauskas *et al.* 1989, Pl. 1:7), suggesting a secondary derivation; the processes of some Svanbergfjellet specimens are markedly constricted where they penetrate the outer membrane (Fig. 18I).

In classifying *Trachyhystrichosphaera* the basic principles of form taxonomy must be adhered to; however, the recognition of its broad intraspecific and taphonomic variation allows some degree of natural taxonomy to be applied. Thus, undetermined or indeterminate features such as process number and morphology (within limits) or the presence/absence of sheath material are not in themselves considered diagnostic characters. Of the 16 published species of *Trachyhystrichosphaera*, 8 appear to be reconcilable to the type species, *T. aimika*, and 7 are misplaced in the genus: '*T.*' *bothnica* and '*T.*' *parva* (unusually small vesicles with questionable processes), '*T.*' *truncata* (solid, evenly distributed processes), and '*T.?*' *anelpa*, '*T.*' *hystricosa*, '*T.*' *triangula* and '*T.*' *tribulosa* (abundant, evenly distributed, homomorphic processes). A new species, *T. polaris* n.sp., is described below.

Trachyhystrichosphaera aimika
Hermann, 1976, emend.

Figs. 18A–K, 19G

Synonymy. – □1969 *Nucellosphaeridium bellum* sp.n. – Timofeev, pp. 23–24, Pl. 6:6 (also Timofeev 1970, Pl. 1C). □1976 *Trachyhystrichosphaera aimika* Hermann sp.n. – Timofeev *et al.*, p. 48, Pls. 19:6, 8; 20:1–3. □1976 *Nucellohystrichosphaera megalea* Timofeev gen. et sp.n. – Timofeev *et al.*, pp. 47–48, Pls. 19:1; 20:4–5. □1978 *Trachyhystrichosphaera aimika* Hermann, 1976 – Jankauskas, p. 915, Pl. 1:1–5. □1980 ?*Trachyhystrichosphaera* sp. – Tynni & Donner, p. 11, Pl. 2:20. □1983 *Trachyhystrichosphaera vidalii* Knoll, 1983 – Knoll & Calder, p. 493, Pl. 58:9–10 □1984 *Trachyhystrichosphaera vidalii* Knoll n.sp. – Knoll, pp. 154–156, Fig. 8A–K. □1985 *Trachyhystrichosphaera aimika* Hermann –

Yin C., Pl. 4:3. □1988 *Trachyhystrichosphaera* – Butterfield & Rainbird, p. A103. □1988 *Trachyhystrichosphaera membranacea* Pjatiletov, sp. nov. – Pjatiletov, pp. 81–82, Pls. 1:1–4; 2:4. □1988 *Trachyhystrichosphaera megalia* (Timofeev, 1976) – Pjatiletov, p. 82, Pls. 1:5; 2:5. □1989 *Trachyhystrichosphaera vidalii* Knoll, 1984 – Allison & Awramik, pp. 287–288, Figs. 11.9, 11.10. □1989 *Trachyhystrichosphaera magna* Allison n.sp. – Allison & Awramik, p. 288, Figs. 11.5–11.8. □1989 *Trachyhystrichosphaera aimica* (*sic*) Hermann, 1976 – Jankauskas *et al.*, p. 46, Pl. 1:6, 8. □1989 *Trachyhystrichosphaera cyathophora* Hermann, sp. nov. – Jankauskas *et al.*, p. 47, Pl. 1:7. □1989 *Trachyhystrichosphaera stricta* Hermann, sp. nov. – Jankauskas *et al.*, p. 47, Pl. 2:4, 7, 8. □1989 *Trachyhystrichosphaera vidalii* Knoll, 1983 – Jankauskas *et al.*, p. 48, Pl. 2:1a–b. □1991 *Trachyhystrichosphaera vidalii* – Knoll *et al.*, Figs. 4.8, 6.4, 7.4–7.8. □1991 *Trachyhystrichosphaera* cf. *magna* – Sergeev, p. 92, Fig. 2. □1993 *Trachyhystrichosphaera vidalii* – Vidal *et al.*, pp. 395–396, Pl. 1:1–4, Fig. 6.

Holotype. – Slide No. 49/2. Laboratoriia biostratigrafii IGGD AN SSSR, Turukhansk collection. Turukhansk Region, Krasnoyarsk District, R. Miroedikha, Miroedikha Formation (Riphean). Timofeev 1969, Pl. 6:6.

Material. – One hundred twenty five (125) specimens: 10 from silicified carbonate samples 86-G-8 and 86-G-15; 115 from shale samples 86-G-62 and P-2945.

Emended diagnosis. – A species of *Trachyhystrichosphaera* with the primary vesicle wall psilate or (taphonomically) shagrinate. Processes 3–8 μm in diameter.

Description. – Psilate or degradationally shagrinate, spheroidal carbonaceous vesicles, 113–702 μm in diameter ($\bar{x}$ = 264 μm; s.d. = 127 μm; n = 123), with one to many hollow, irregularly distributed processes, 3–8 μm in maximum diameter. Processes communicate freely with the vesicle but are otherwise highly heteromorphic: conical, tubular, capitate, basally bifurcate and variously expanded or constricted forms; up to 45 μm long. Sheath usually present but of variable character, ranging from inconspicuous surface layers to well defined and fully encompassing envelopes. Silicified specimens with a thick, relatively dense inner envelope and a thin, diffluent and apparently tacky outer layer.

Discussion. – *Trachyhystrichosphaera aimika* is readily identified by its variable process morphology and thin psilate walls. A sheath commonly surrounds the vesicle, and close examination of silicified specimens shows it to be bilayered. The tacky outer layer apparently served to attach the vesicle to solid substrates (Fig. 18K), much like that seen in *Cymatiosphaeroides kullingii*; both would appear to have had an at least intermittently benthic habit.

Silicified *T. aimika* in the Svanbergfjellet Formation differ only superficially from the more common shale-hosted material and show a closely comparable size-frequency distribution ($\bar{x} = 265$ µm; s.d. $= 137$ µm; $n = 10$). More difficult to identify are forms with very short or eroded processes (Fig. 18A, G); in the latter case, specimens may only be identifiable by the distinctively irregular arrangement of perforations left in a vesicle wall (Fig. 13A). The morphological range of *T. aimika* is further extended by vesicles with marked, more or less medial constrictions (Fig. 18G), and a single specimen showing evidence of binary fission; two *T. aimika* within a common envelope have been recorded in the overlying Draken Conglomerate Formation (A.H. Knoll, unpublished observation, 1991).

Close examination of the type specimen of *Nucellosphaeridium bellum* Timofeev, 1969 (Timofeev 1969, Pl. 6:6; 1970, Pl. 1C), from the Miroedikha Formation shows it to be conspecific with *T. aimika*. To accept 'bellum' as the senior synonym, however, is untenable as it would entail the changing of types (see also ICBN Article 43.1). *Trachyhystrichosphaera aimika* is thus retained as the type, and, in our opinion, the only currently legitimate species of the genus.

Trachyhystrichosphaera aimika has been reported from a number of Siberian sequences, Kazakhstan, the Southern Urals, Finland, China, various formations on Spitsbergen, and from the Yukon Territory and Northwest Territories of Canada. In all instances it can be reliably constrained to the Late Riphean, making it a valuable index fossil for this time period.

Trachyhystrichosphaera polaris Butterfield, n.sp.

Fig. 19A–F, H

Holotype. – HUPC 62717, Fig. 19E; Slide 86-G-62-1M, England-Finder coordinates U-30-3.

Type locality. – Algal Dolomite Member, Svanbergfjellet Formation, Geerabukta (79°35'30"N, 17°44'E); 55 m above base of member.

Material. – Twenty one (21) specimens from shale sample 86-G-62. Seven (7) designated paratypes: HUPC 62817–62821, Fig. 19A–D, 19F; HUPC 62823, Fig. 19H.

Diagnosis. – A species of *Trachyhystrichosphaera* with an echinate primary vesicle. Processes 3–5 µm in diameter.

Description. – Echinate, spheroidal, dark carbonaceous vesicles, 95–235 µm in diameter ($\bar{x} = 160$ µm; s.d. $= 33$ µm; $n = 20$) with a variable number of randomly distributed processes, 3–5 µm in diameter. Echinate structures solid and closely packed: thin and hair-like on small specimens, thick (ca. 1 µm) and straight on larger ones. Processes hollow and heteromorphic: tubular, evexate, basally constricted and/or terminally flared; up to 55 µm long and sometimes confluent with an outer envelope. Envelope present or absent.

Discussion. – The processes of *T. polaris* n.sp. show very much the same range of morphology and distribution as those of *T. aimika*. Similarly, an external envelope is usually, but not necessarily (Fig. 19A–C, F) present, and the style of intrapopulational (ontogenetic?) variation appears to be closely comparable. The only substantial difference between the two is the pronounced echinate surface texture of *T. polaris*, which is here taken as a species-level distinction. Subsidiary differences include a more restricted size range for *T. polaris*, its somewhat darker walls, and an often more clearly delineated envelope than is typical for *T. aimika*; comparable envelopes are found in the type species, however (Fig. 19G; Jankauskas *et al.* 1989, Pl. 2:1).

The Svanbergfjellet population of *T. polaris* n.sp. includes relatively small and thin-walled unsheathed vesicles with folds (Fig. 19A, B), and larger, unfolded and almost opaque forms, with or without envelopes. The specimens in Fig. 19C and F appear never to have had envelopes, but they clearly belong to the same taxon as the ensheathed examples in Fig. 19E and H. The specimen in Fig. 19D also lacks an envelope but in this case it appears to have lost it through erosion; remnants have remained attached to the terminally flared processes. If these specimens represent in any sense an ontogenetic series, it would appear to progress from Fig. 19C to E to D.

Etymology. – With reference to its northerly occurrence and the star-like appearance of the type specimen.

Domain Bacteria Woese, Kandler & Wheelis, 1990

Kingdom Eubacteria Woese & Fox, 1977

Phylum Cyanobacteria Stanier *et al.*, 1978

Class Coccogoneae Thuret, 1875

Order Chroococcales Wettstein, 1924

Family Chroococcaceae Nägeli, 1849

Genus *Gloeodiniopsis* Schopf, 1968, emend. Knoll & Golubic, 1979

Type species. – *Gloeodiniopsis lamellosa* Schopf, 1968, p. 684.

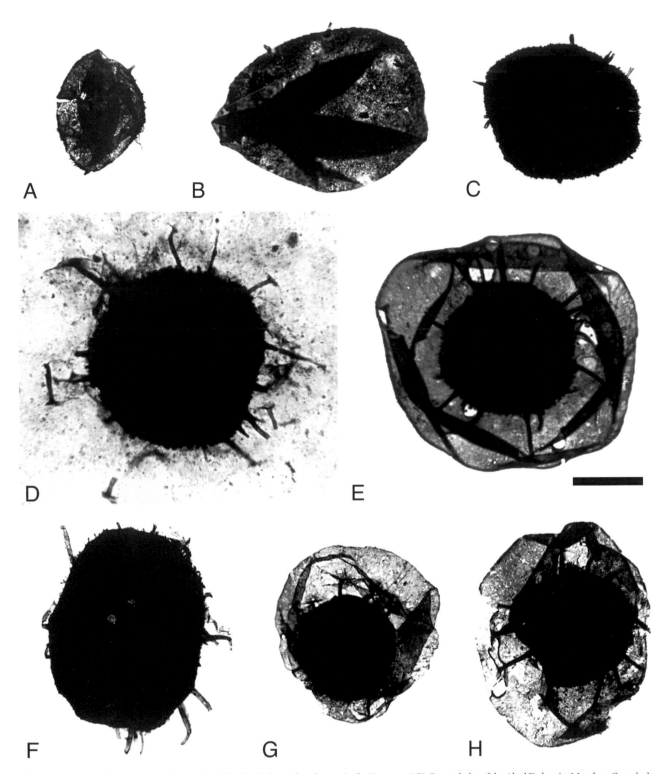

Fig. 19. Trachyhystrichosphaera polaris n.sp. (A–F, H); *Trachyhystrichosphaera aimika* Hermann (G). From shales of the Algal Dolomite Member, Geerabukta. Scale bar in E equals 50 μm for A, B, D; 80 μm for C, E–H. □A. HUPC 627817; 86-G-62-11M (Q-28-0); with hair-like surface ornamentation. □B. HUPC 62818; 86-G-62-168M (O-32-0). □C. HUPC 62819; 86-G-62-205M (N-33-2); with heteromorphic processes. □D. HUPC 62820; 86-G-62-40 (O-24-3); in bedding-parallel thin-section; with remnants of the envelope at process termini. □E. HUPC 62717; 86-G-62-1M (U-30-3); holotype. □F. HUPC 62821; 86-G-62-206M (P-34-2). □G. HUPC 62822; 86-G-62-102M (K-33-0); with smooth inner vesicle. □H. HUPC 62823; 86-G-62-101M (H-37-0).

Fig. 20 (opposite page). *Eoentophysalis croxfordii* (Muir) n.comb. (A); *Eoentophysalis* sp. (B); *Myxococcoides minor* Schopf (C); *Eoentophysalis belcherensis* Hofmann (D, E); *Myxococcoides cantabrigiensis* Knoll (F, J); *Myxococcoides* sp. (G); *Gloeodiniopsis lamellosa* Schopf. (H); *Chlorogloeaopsis zairensis* Maithy (I); *Sphaerophycus* sp. (K); *Sphaerophycus parvum* Schopf (L–T). From cherts (A, D–H) and shales (I) of the Algal Dolomite Member, Geerabukta; cherts of the Lower Limestone Member, Polarisbreen (K); and cherts of the Lower Dolomite Member, Polarisbreen (C, J, L–T). Scale bar in T equals 50 μm for A, D, F, T;

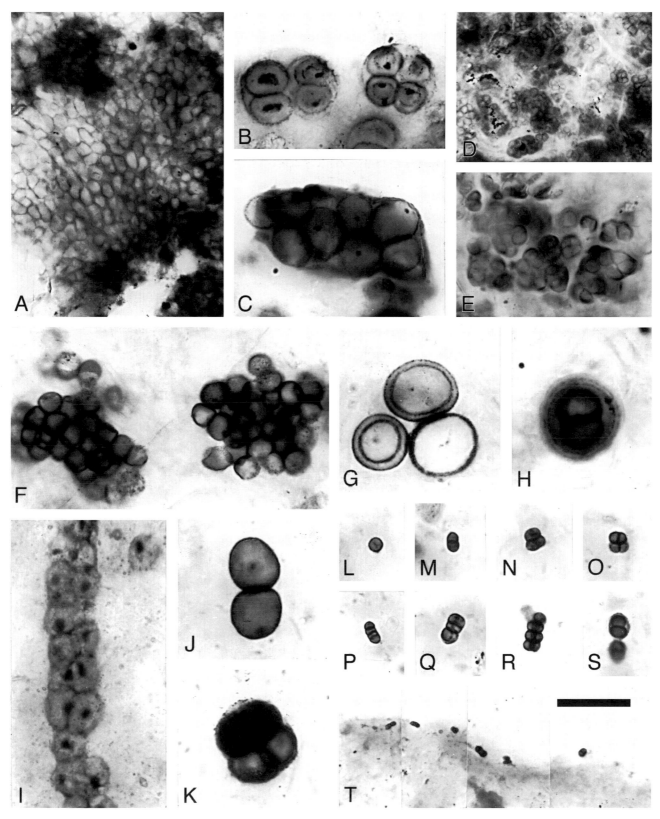

20 μm for B, C, E, G–S. □A. HUPC 62824; 86-G-15-1 (E-59-2). □B. misplaced. □C. HUPC 62825; P-2664-4A (J-55-4). □D. HUPC 62826; 86-G-8-2B (E-61-3). □E. HUPC 62864; 86-G-4-1A (A-54-3). □F. HUPC 62827; 86-G-14-1A (P-49-4). □G. HUPC 62828; 86-G-14-1A (T-72-1). □H. HUPC 62829; 86-G-9-2C (C-56-1). □I. HUPC 62830; 86-G-62-44 (N-50-3). □J. HUPC 62831; P-2664-4A (D-58-3). □K. HUPC 62832; P-3085-1B (L-52-0). □L. HUPC 62833; 86-SP-8-1 (R-60-4). □ M. HUPC 62834; 86-SP-8-1 (M-60-2). □ N. HUPC 62835; 86-SP-8-1 (G-43-2). □O. HUPC 62836; 86-SP-8-1 (K-55-4). □P. HUPC 62837; 86-SP-8-1 (P-62-4). □Q. HUPC 62838; 86-SP-8-1 (K-55-4). □R. HUPC 62839; 86-SP-8-1 (K-54-4). □S. HUPC 62840; 86-SP-8-1 (H-54-0); with sheath. □T. 86-SP-8-1 (P-62-4); population of *S. parvum* colonizing the surface of a microbialite intraclast.

Gloeodiniopsis lamellosa Schopf, 1968, emend. Knoll & Golubic, 1979

Fig. 20H

Material. – Two (2) specimens from chert samples 86-G-8 and 86-G-9.

Discussion. – Knoll *et al.* (1991) considered as *Gloeodiniopsis* only those populations of spheroidal microfossils that exhibit *both* multiple envelopes and 2–8 daughter cells within a common envelope. Spheroids with more than one envelope in the Svanbergfjellet assemblage occur intermittently in low diversity microbial-mat intraclasts; however, multiple cells with multiple envelopes are rare. Of the latter, one specimen is associated with a large population of non-enveloped or single-enveloped *Myxococcoides*, suggesting some genetic or ontogenetic relationship between the two.

Family Entophysalidaceae Geitler, 1925

Genus *Eoentophysalis* Hofmann, 1976, emend. Mendelson & Schopf, 1982

Type species. – *Eoentophysalis belcherensis* Hofmann, 1976, pp. 1070–1072.

Eoentophysalis belcherensis Hofmann, 1976

Fig. 20D–E

Synonymy. – □1976 *Eoentophysalis belcherensis* Hofmann n.sp. – Hofmann, pp. 1070–1072, Pls. 4:1–5; 5:3–6; 6:1–14. □1976 *Myxococcoides kingii* sp. nov. – Muir, pp. 151–152, Fig. 6H. □1979 *Eoentophysalis cumulus* n.sp. – Knoll & Golubic, pp. 148–149, Figs. 2E, 3.

Material. – Scattered colonies in chert samples 86-G-4 and P-2664.

Description. – Spheroidal vesicles, 3–5 μm in diameter, arranged in small pustulose clusters or extensive colonies of several hundred vesicles. Diads and tetrads commonly isolated by an enveloping sheath.

Discussion. – Except for the luxuriant mat-forming habit, Svanbergfjellet *E. belcherensis* exhibit most of the variation found in the type populations. Furthermore, they show no clear morphological break between small isolated colonies (Fig. 20E) and incipient mats (Fig. 20D). Golovenoc & Belova (1984) have suggested that the more isolated colonial forms be assigned to a separate genus, *Eogloeocapsa*; however, the type specimen of *Eoentophysalis* is already of this particular habit (Hofmann 1976, Pl. 6:13), making such a designation untenable. *Eoentophysalis cumulus* differs only

in age from *E. belcherensis* (Sergeev & Krylov 1986) and is thus its junior synonym, as is *Myxococcoides kingii*.

Eoentophysalis croxfordii (Muir, 1976) Butterfield, n.comb.

Fig. 20A

Synonymy. – □1976 *Ameliaphycus croxfordii* sp. nov. – Muir, p. 155, Figs. 6L–M, 7A–B.

Holotype. – CPC15603, 115.0:14.4; Amelia Dolomite; Muir 1976, Fig. 6M.

Material. – Six (6) occurrences in chert samples 86-G-9, 86-G-14 and 86-G-15.

Diagnosis. – A species of *Eoentophysalis* with vesicles 8–16 μm in diameter.

Description. – Massive pluricellular aggregations of close-packed, often compressionally distorted spheroidal vesicles; usually separated from one another by a substantial (ca. 2 μm thick) extracellular layer. Vesicles equidimensional to slightly elongate, 8–16 μm in diameter ($\bar{x}$ ca. 12 μm).

Discussion. – A number of intraclasts in the Lower Dolomite Member microbialite grainstones are constructed exclusively of *E. croxfordii* and, with dimensions of up to 1.5 mm, can comprise several thousand individual vesicles. Mutual compression has resulted in a typically polygonal cellular outline, although the vesicles themselves are usually separated by a relatively thick mucilaginous(?) envelope. This intervening layer is sometimes absent between paired smaller cells, suggesting that growth occurred by way of binary fission.

The fragmentary nature of the present material precludes a positive determination of overall colony morphology; however, the occasional suggestion of a radial fabric (Fig. 20A) may be reflective of mammillate microbial-mat formation, such as that reported for modern *Entophysalis* (cf. Golubic & Hofmann 1976). Alternatively, the fragments may derive from more complex pluri- or multicellular structures such as *Thallophyca* Zhang, 1989; some higher order organization is suggested in several of the Svanbergfjellet specimens where cells surround and define what appear to have been internal spaces.

Eoentophysalis croxfordii n.comb. differs from the mat-forming variant of *E. belcherensis* primarily in the considerably larger dimensions of its cells, those of the type species having a mean diameter of just 3.9 μm. *Eoentophysalis yudomatica* Lo, 1980, and *E. arcata* Mendelson & Schopf, 1982 (Fig. 20B), have more comparable cell dimensions (ca. 12 μm) but fail to express a dense colonial/mat-forming habit. *Eoentophysalis dismallakesensis* Horodyski & Donaldson, 1980, and *E. magna* McMenamin *et al.*, 1983, lack both the size and the habit of *E. croxfordii*.

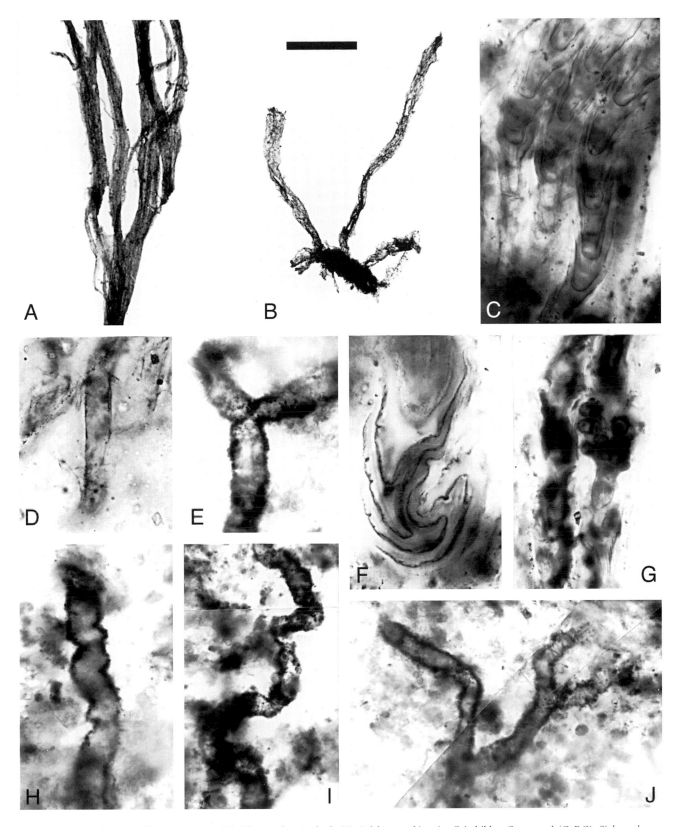

Fig. 21. Pseudodendron anteridium n.gen., n.sp. (A); Filament-bearing body (B); *Polybessurus bipartitus* Fairchild *ex* Green *et al.* (C, F,G); *Siphonophycus kestron* Schopf (D); *Palaeosiphonella* sp. (E, H–J). From shales of the Algal Dolomite Member, Geerabukta (A, B); cherts of the Lower Limestone Member, Polarisbreen (C, E–J); and cherts of the Lower Dolomite Member (D). Scale bar in B equals 75 μm for A, G, I, J; 120 μm for B, C; 50 μm for D–F, H. □A. HUPC 62721; 86-G-62-202M (O-42-3); holotype. □B. HUPC 62841; 86-G-62-70M (L-27-0). □C. HUPC 62842; 86-P-89-1 (H-43-4). □D. HUPC 62843; P-2628-1A (V-48-1). □E. HUPC 62844; P-3075-1F (S-47-3). □ F. HUPC 62845; 86-P-89-1C (U-37-4). □G. HUPC 62846; 86-P-89-1C (T-68-3); with cluster of probable baeocytes. □H. HUPC 62847; P-3075-1A (Z-55-3). □I. HUPC 62848; P-3075-1C (N-60-4). □J. HUPC 62849; P-3075-1A (Z-55-4).

Order Pleurocapsales Geitler, 1925

Family Dermocarpaceae Geitler, 1925

Genus *Polybessurus* Fairchild, 1975, *ex* Green *et al.*, 1987

Type species. – *Polybessurus bipartitus* Fairchild, 1975, *ex* Green *et al.*, 1987, p. 938.

Polybessurus bipartitus Fairchild, 1975, *ex* Green *et al.*, 1987
Fig. 21C, F–G

Material. – Massive population in chert sample 86-P-89; also in P-3400.

Discussion. – This distinctive stalk forming cyanobacterium is found in the Lower Limestone Member where it occurs both in small isolated colonies associated with abundant *Siphonophycus typicum* n.comb., and as massive crust-forming accumulations. In the latter case it occupies, almost to the exclusion of other taxa, a ca. 2 cm high by 5 cm wide mound; the remaining diversity in this crust is limited to localized populations of *S. typicum* and various spheroids.

The paleobiology of *P. bipartitus* was discussed in detail by Green *et al.* (1987). Its pattern of cell division (by baeocyte production), and thereby its ordinal-level classification, was inferred from the geometrical arrangement of close-packed stalks. The Svanbergfjellet material supports these previous interpretations and possibly offers direct confirmation of the reproductive mode. Midway up and within the *Polybessurus* stalk shown in Fig. 21G is a cluster of ca. 12-μm-diameter spheroids, much smaller than the cell that would have formed the stalk. These cells likely represent the baeocytes (cell fission without intervening growth) inferred by Green *et al.* (1987, p. 935). Branching of *Polybessurus* stalks was also considered possible but not observed in the East Greenland fossils (Green *et al.* 1987); it is rarely encountered in the Svanbergfjellet populations.

Polybessurus is proving to have had both wide geographical and temporal distribution in the Proterozoic. It has now been reported from two ca. 725–1250 Ma sequences from the Canadian Arctic (Butterfield *et al.* 1990; Hofmann & Jackson 1991), the Middle Riphean of the Southern Urals (Sergeev 1991), 750–850 Ma sections in South Australia (Fairchild 1975), and in 700–800 Ma old rocks from East Greenland and Spitsbergen (Green *et al.* 1987; Knoll *et al.* 1991; present study).

Class Hormogoneae Thuret, 1875

Order Oscillatoriales Copeland, 1936

Family Oscillatoriaceae Kirchner, 1898

Genus *Obruchevella* Reitlinger, 1948, emend. Knoll, 1992

Type species. – *Obruchevella delicata* Reitlinger, 1948, p. 78.

Discussion. – The record of helically coiled microfossils was recently reviewed by Knoll (1992) and Mankiewicz (1992), with the former recommending that all such forms deriving from intact extracellular sheaths (as opposed to split sheaths or cellular trichomes) be placed in the form genus *Obruchevella*. The artificial distinction between three-dimensional (*Obruchevella*) and two-dimensional preservation (*Volyniella*) was earlier recognized by Jankauskas *et al.* (1989).

Obruchevella blandita Schenfil, 1980
Fig. 22A–F

Synonymy. – ☐1980 *Obruchevella blandita* Schenfil sp.n. – Schenfil, p. 994, Pl. 3. ☐1980 *Volyniella glomerata* Jankauskas, sp. nov. – Jankauskas, 1980b, p. 112, Pl. 12:18–19. ☐1984 *Obrachevella* [*sic*] *condensata* Liu sp. nov. – Liu *et al.*, p. 177, Pl. 1:10. ☐1989 *Glomovertella glomerata* (Jankauskas, 1980) Jankauskas, comb. nov. – Jankauskas *et al.*, pp. 108–109, Pl. 31:8–10.

Material. – Several hundred specimens from shale samples 86-G-28, 86-G-30, 86-G-61 and 86-G-62.

Description. – Nonseptate hollow filament, wound into a regular helix with adjacent coils in close contact (pitch = filament diameter). Filament diameter constant at 1.5 μm. Helix diameter 6–27 μm ($\bar{x}$ = 15.9 μm; s.d. = 3.9 μm; n = 150); usually uniform in a single specimen, but sometimes contracted/expanded by a factor of two over a translational distance of 10 μm; occasional thick, dark annulations.

Discussion. – *Obruchevella blandita* was originally described from silicified, three-dimensionally preserved material from the Upper Riphean of the Yenisey Ridge region, Siberia. Its two-dimensional counterpart is common in Svanbergfjellet shales where it occurs as intact units up to 100 μm long (Fig. 22A), as disaggregating and partially disaggregated fragments (Fig. 22B–D, F), and as isolated ring structures (Fig. 22E). The Svanbergfjellet population differs from the type material in its slightly narrower filament diameter (1.5 μm *vs.* 2.1–2.2 μm), greater range of helix diameter (6–27 μm *vs.* 18–20 μm), and by the occasional occurrence of thick external annulations (Fig. 22C, F); these features are here considered as reasonably accommodated by *O. blandita*, the latter

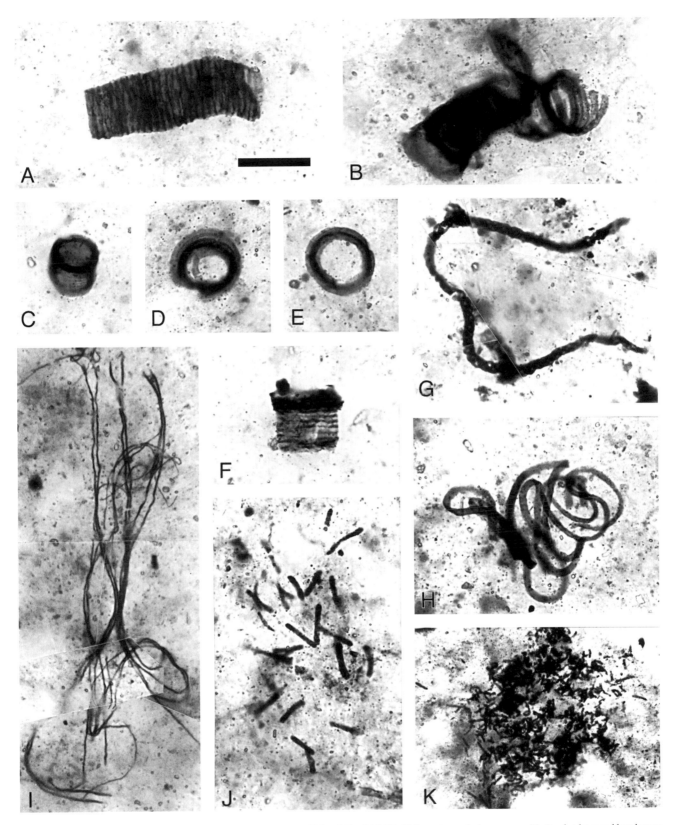

Fig. 22. Obruchevella blandita Schenfil (A–F); *Siphonophycus septatum* (Schopf) Knoll (G, H); *Siphonophycus thulenema* n.sp. (I); *Brachypleganon khandanum* Lo (J, K). From shales of the Algal Dolomite Member, Geerabukta. Scale in A equals 20 μm for A–J; 50 μm for K. □A. HUPC 62850; 86-G-62-1 (P-48-2); with tubular cross-section on the exposed terminal coil. □B. HUPC 62851; 86-G-30-4 (N-27-4); disaggregating specimen. □C. HUPC 62852; 86-G-28-2 (N-27-3). □D. HUPC 62853; 86-G-62-16 (L-18-1); tranverse cross-section. □E. HUPC 62854; 86-G-62-11 (N-22-4); isolated coil. □F. HUPC 62855; 86-G-61-4 (S-44-1); with annular thickening. □G. HUPC 62856; 86-G-62-24 (K-37-3); entwined filaments. □H. HUPC 62857; 86-G-62-24 (P-47-1). □I. HUPC 62718; 86-G-62-46 (R-28-3); holotype. □J. HUPC 62858; 86-G-62-47 (N-50-4). □K. HUPC 62859; 86-G-62-10 (R-15-2); aggregated and loosely oriented colony.

apparently the result of one or more outer, superimposed coils (cf. Mankiewicz 1992). Unravelling of the helices is also common in the Svanbergfjellet population and in this habit is indistinguishable from conspecific *Glomovertella glomerata*. Unravelling also occurs in the type species, *O. delicata* (Mankiewicz 1992).

Phylum Cyanobacteria(?) Stanier *et al.*, 1978

Class Coccogoneae(?) Thuret, 1875

Order Chroococcales(?) Wettstein, 1924

Family Chroococcaceae(?) Nägeli, 1849

Genus *Eosynechococcus* Hofmann, 1976, emend. Golovenoc & Belova, 1984

Synonymy. – □1976 *Eosynechococcus* n.gen. – Hofmann, pp. 1069–1070. □*non* 1984 *Eosynechococcus elongatus* V. Golovenoc et M. Belova, sp. nov. – Golovenoc & Belova, p. 29, Pl. 2:5.

Type species. – *Eosynechococcus moorei* Hofmann, 1976, pp. 1057–1058.

Discussion. – *Eosynechococcus* was originally erected for ellipsoidal microfossils with a length-to-width ratio of ca. 2:1 (Hofmann 1976). Subsequently named species (excluding *E. elongatus*) increased this ratio to 3.6:1 (Golovenoc & Belova 1984, Table 2), which is here taken as the upper limit for the genus. The asymmetrical cell division(?) and markedly longer aspect ratio of *E. elongatus* (6:1) place it well outside reasonable limits for the genus.

Eosynechococcus moorei Hofmann, 1976
Fig. 23J

Material. – Several small populations in chert sample P-2664.

Description. – Ellipsoidal vesicles approximately twice as long as wide (ca. 6×3 μm) and typically linked end to end; outer envelope absent.

Genus *Sphaerophycus* Schopf, 1968

Type species. – *Sphaerophycus parvum* Schopf, 1968, p. 672.

Sphaerophycus parvum Schopf, 1968
Fig. 20L–T

Material. – Several hundred specimens in chert sample 86-SP-8; others in P-2664 and 86-G-15.

Description. – Spheroidal vesicles, 2.0–3.5 μm in diameter, usually without enveloping sheaths; solitary or grouped into small regular colonies of 2–16 cells.

Discussion. – Svanbergfjellet *S. parvum* closely approximate the size and structure of the Bitter Springs type material, the only significant difference being the tighter, more regular packing of their pluricellular colonies. In one sample (86-SP-8) they also exhibit an interesting behavioral aspect, occurring on and along the exposed surfaces of intraclastic grains (Fig. 20T). Such distribution indicates that they had actively colonized the recently exposed surfaces, while a geopetal preference for one surface (presumably the upper) supports their interpretation as photosynthesizers. These epilithic populations are almost certainly related to the less conspicuous *S. parvum* that occur within the intraclasts of the same sample.

Sphaerophycus spp.
Fig. 20K

Discussion. – Colonial fossils comparable to named species of *Sphaerophycus*, but of considerably larger dimensions, are occasionally encountered in Svanbergfjellet cherts. A distinctive population of spheroids, ca. 16 μm in diameter, occurring as unicells, diads, and tetrads is closely associated with crust-forming *Polybessurus bipartitus* in sample 86-P-89.

Class Hormogoneae(?) Thuret, 1875

Order Oscillatoriales(?) Copeland, 1936

Family Oscillatoriaceae(?) Kirchner, 1898

Genus *Cephalonyx* Weiss, 1984

Synonymy. – □1980 *Calyptothrix* Schopf, 1968 – Jankauskas 1980b, pp. 107–108; *non* 1968 *Calyptothrix* Schopf, n.gen. – Schopf, pp. 667–669. □1983 *Striatella* Assejeva gen. nov. – Assejeva & Velikanov, p. 6; *non* 1964 *Striatella* Mädler gen. nov. – Mädler, p. 189. □1984 *Cephalonyx* A. Weiss, gen.

Fig. 23. Cephalonyx geminatus (Jankauskas) n.comb. (A); *Siphonophycus*-like filaments with localized encrustations (B–D); *Oscillatoriopsis amadeus* (Schopf & Blacic) n.comb. (E); Sub-vertical branched(?) filaments in shale-hosted apatite nodules (F, G); *Tawuia dalensis* Hofmann, SEM of a broken edge (H); *Pseudodendron*(?) sp. (I); *Eosynechococcus moorei* Hofmann (J); *Leiosphaeridia crassa* (Naumova) Jankauskas, in dolomitic intraclast (K). From shales (A, E, H, I) and cherts (B–D) of the Algal Dolomite Member, Geerabukta; shale-hosted apatite of the Lower Dolomite Member, Svanbergfjellet (F, G); cherts of the

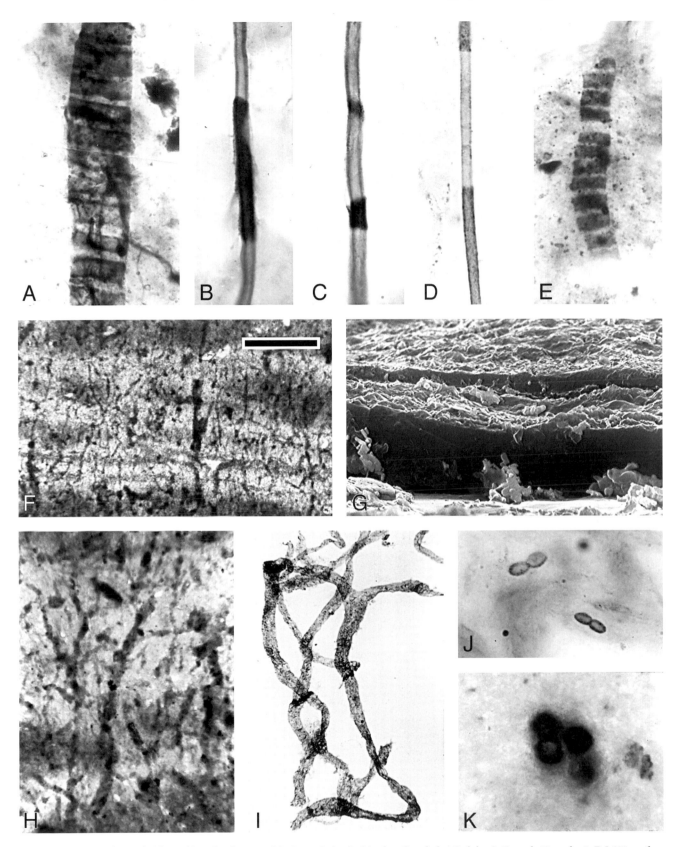

Lower Dolomite Member, Polarisbreen (J); and carbonates of the Lower Dolomite Member, Geerabukta. Scale bar in F equals 20 µm for A–E, J; 375 µm for F; 185 µm for G; 75 µm for I; 50 µm for K; 3.5 µm for H. □A. HUPC 62860; 86-G-62-17 (M-45-4). □B. HUPC 62861; 86-G-15-2A (Z-30Ø). □C. HUPC 62861; 86-15-2A (Z-30Ø). □D. HUPC 62862; 86-G-15-3 (O-49-1). □E. HUPC 62863; 86-G-62-65 (N-30-3). □F. HUPC 61374; 86-SV-3-1A (O-41-2). □G. HUPC 61374; 86-SV-3-1A (R-42-1); detail of F. □H. HUPC 62865; 86-G-62-2S; SEM. □I. HUPC 62866; 86-G-62-94M (M-34-0). □J. HUPC 62867; P-2664-2A (P-67-0). □K. HUPC 62868; 86-G-9-3 (S-56-2).

nov. – Weiss, pp. 105–106. ☐1984 *Arthrosiphon* A. Weiss gen. nov. – Weiss, p. 106. ☐1985 *Contextuopsis* Hermann, gen. nov. – Sokolov & Ivanovskij, p. 150. ☐1989 *Rectia* Jankauskas, gen. nov. – Jankauskas *et al.*, pp. 120–121.

Type species. – *Cephalonyx coriaceus* (Assejeva, 1983) n.comb., p. 6 (*in* Assejeva & Velikanov 1983).

Diagnosis. – Unbranched filamentous sheaths with annular thickenings that reflect the original positioning of cells (pseudocellular filaments). Annulations typically of greater diameter than the intervening regions of the sheath.

Discussion. – Pseudocellular fossil filaments were first described by Assejeva (1983) and given the binomial *Striatella coriacea*; however, an earlier homonym (Mädler 1964) renders this generic name illegitimate (ICBN Article 64.1). The earliest legitimate generic name (ICBN Article 57.1) that can be applied to these fossils is *Cephalonyx* Weiss, 1984, a pseudocellular filament from the Late Riphean Miroedikha Suite, Siberia. Consequently, the type species is reconstituted as *Cephalonyx coriaceus* (Assejeva, 1983).

Pseudocellular *Cephalonyx* is potentially confused with *Palaeolyngbya*; however, the regular annulations of *Cephalonyx* are simply external(?) thickenings in a filamentous sheath, not true cellular remains (although they do appear to reflect the original positioning of cells; Jankauskas 1980b). By contrast, the annulations in pseudoseptate *Tortunema* represent the original positioning of intercellular *septa*, while the transverse fabric of the outer sheath of *Rugosoopsis* is entirely independent of the trichome.

Cephalonyx geminatus (Jankauskas, 1980) Butterfield, n.comb.

Fig. 23A

Synonymy. – ☐1980 *Calyptothrix geminata* Jankauskas, sp. nov. – Jankauskas 1980b, pp. 107–108, Pl. 12:20.

Holotype. – LitNIGRI No 16-4-3526/6, spec. 2; Ural forelands, Bashkiria; Borehole Kabakovo-62, interval 3526–3528 m; Upper Riphean. Jankauskas 1980b, Pl. 12:20.

Material. – One (1) specimen from shale sample 86-G-62.

Diagnosis. – A species of *Cephalonyx* with filamentous sheaths, 10–20 μm in diameter.

Description. – Filamentous sheath, 17 μm wide, with prominent, 4–5 μm long annulations separated by 1–2 μm long thin-walled intervals. Pseudocellular annulations with marginally greater diameter than intervening areas. Intact terminus simply rounded.

Discussion. – Reexamination of the type material of *Calyptothrix* Schopf, 1968, shows it to be a true cellular trichome that, at least in places, retains a relatively thick mucilaginous

sheath. It is clearly unrelated to pseudocellular *Cephalonyx*, including '*Calyptothrix*' *geminata*. The single specimen of Svanbergfjellet *C. geminatus* n.comb. is marginally wider than the size range suggested by the original description (17 μm *vs.* 13–15 μm), but it clearly shows the same arrangement of pseudocellular thickenings and the thin interannular regions of the holotype.

Genus *Cyanonema* Schopf, 1968, emend.

Synonymy. – ☐1968 *Cyanonema* Schopf, n.gen. – Schopf, p. 670. ☐*non* 1987 *Cyanonema disjuncta* Ogurtsova et Sergeev, sp. nov. – Ogurtsova & Sergeev, pp. 111–112, Pl. 9:3–4 (= *Oscillatoriopsis*).

Type species. – *Cyanonema attenuata* Schopf, 1968, p. 670.

Emended diagnosis. – Unbranched uniseriate cellular trichomes in which cell length is greater than cell diameter (length:width >1). Sheath absent. Not at all to moderately constricted at septa.

Discussion. – The distinction between 'cell length ≤ diameter' (= *Oscillatoriopsis*) *vs.* 'cell length > diameter' (= *Cyanonema*) is certain to be artificial to some extent; it nevertheless serves as an unambiguous character in form taxonomy. Cell length-to-width ratios are similarly applied in the botanical classification of modern cyanobacteria (e.g., Geitler 1925).

In addition to the type, four species of *Cyanonema* have appeared in the literature: *C. disjuncta* has cells shorter than wide and therefore belongs in *Oscillatoriopsis*, while the others, *C. inflatum* Oehler, 1977, *C. minor* Oehler, 1977, and *C. ligamen* Zhang, 1981, would appear to be accommodated by *C. attenuata*.

Cyanonema sp.

Fig. 24J

Material. – One (1) specimen from shale sample 86-G-62.

Discussion. – A single trichome in the upper Svanbergfjellet shales has cells ca. 7 μm wide by 12 μm long and is therefore classified as *Cyanonema*. Its relatively large dimensions exclude it from *C. attenuata*.

Genus *Oscillatoriopsis* Schopf, 1968, emend.

Synonymy. – ☐1968 *Oscillatoriopsis* Schopf, n. gen. – Schopf, p. 666. ☐1968 *Halythrix* Schopf, n.gen. – Schopf, p. 678. ☐*non* 1978 *Oscillatoriopsis*? *hubeiensis* Yin et Li (sp. nov.) – Yin & Li, pp. 88–89, Pl. 7:9. ☐*nec* 1980 *Oscillatoriopsis robusta* n.sp. – Horodyski & Donaldson, pp. 149–152, Fig. 13H.

(= *Palaeolyngbya*). □*nec* 1980 *Oscillatoriopsis curta* n. sp. – Horodyski & Donaldson, p. 149, Fig. 13A–G (= *Cyanonema*). □*nec* 1980 *Oscillatoriopsis bothnica* n.sp. – Tynni & Donner, p. 15, Pl. 7:83 (= *Tortunema*). □*nec* 1980 *Oscillatoriopsis constricta* n.sp. – Tynni & Donner, p. 15, Pl. 7:82, 85, 86. (= *Tortunema*). □*nec* 1980 *Oscillatoriopsis magna* n.sp. – Tynni & Donner, p. 14, Pl. 6:64–66, 68–70. (= *Tortunema*). □*nec* 1981 *Oscillatoriopsis bacillaris* Hermann, sp. nov. – Hermann 1981b, p. 121, Pl. 12:13 (= *Tortunema*). □*nec* 1981 *Oscillatoriopsis nochtuica* Yakschin, sp. nov. – Yakshin & Luchinina, p. 31, Pl. 11:3a–b. □*nec* 1981 *Oscillatoriopsis tomica* Yakschin, sp. nov. – Yakshin & Luchinina, p.32, Pl. 12:1, 2, 5. □*nec* 1983 *Oscillatoriopsis platensis* Zhang, sp. nov. – Zhang, pp. 213, 219, Pl. 2:5. □*nec* 1984 *Oscillatoriopsis acuminata* Xu sp. nov. – Xu, pp. 218, 312, Pl. 1:3, 4, 6 (= *Siphonophycus*). □*nec* 1984 *Oscillatoriopsis disciformis* Xu sp. nov. – Xu, pp. 218–219, 313, Pl. 3:7 (= *Siphonophycus*). □*nec* 1984 *Oscillatoriopsis glabra* Xu sp. nov. – Xu, pp. 219, 313, Pls. 2:6, 8A; 3:9, 11 (= *Siphonophycus*). □*nec* 1984 *Oscillatoriopsis hemisphaerica* Xu sp. nov. – Xu, pp. 218, 312, Pls. 1:7–8; 2:12 (= *Siphonophycus*). □*nec* 1984 *Oscillatoriopsis tuberculata* Xu sp. nov. – Xu, pp. 219, 313, Pl. 1:1 2 (= *Siphonophycus*). □*nec* 1985 *Oscillatoriopsis funiformis* Ragozina, sp. nov. – Sokolov & Ivanovskij pp. 142–143, Pl. 59:2–3. □*nec* 1985 *Oscillatoriopsis rhomboidalis* Siverzeva, sp. nov. – Sokolov & Ivanovskij, p. 143, Pl. 60:4, 6, 7 (= *Pomoria*). □*nec* 1985 *Oscillatoriopsis minuta* sp. nov. – Lu & Zhu, p. 84, Pl. 1:3, 5, 6. □*nec* 1985 *Oscillatorjopsis* [*sic*] *tenuis* sp. nov. – Lu & Zhu, p. 84, Pl. 1:4. □*nec* 1986 *Oscillatoriopsis hechiensis* Lie et Liu sp. nov. – Liu & Li, pp. 266, 270, Pl. 1:2. □*nec* 1989 *Oscillatoriopsis subtilis* sp. nov. – Zhang *et al.*, pp. 323–324, Pl. 1:3, 4. (= *Cyanonema*). □*nec* 1989 *Oscillatoriopsis angusta* (Kolosov, 1984), comb. nov. – Jankauskas *et al.*, p. 116 (= *Tortunema*).

Type species. – *Oscillatoriopsis obtusa* Schopf, 1968, p. 667.

Emended diagnosis. – Unbranched, uniseriate cellular trichomes with cell length less than or equal to cell diameter (length:width ≤1). Sheath absent. Not at all to moderately constricted at septa.

Discussion. – *Oscillatoriopsis* is here emended to include only those filamentous microfossils that can be confidently identified as cellular trichomes (*vs.* pseudoseptate sheaths – see *Tortunema*), lack an enveloping sheath (*vs.* sheathed – see *Cephalonyx, Palaeolyngbya, Rugosoopsis*), have a cell length:width ratio less than or equal to 1 (*vs.* >1 – see *Cyanonema*), and are not severely constricted at intercellular septa (*vs.* severely constricted – see *Veteronostocale*), all these being characteristics of the type species. Through failing to satisfy one or more of these criteria, all the above listed species are hereby removed from *Oscillatoriopsis*.

Given the marked variation recorded in terminal-cell morphology, trichome length and septal constriction of both living and moribund oscillatorian cyanobacteria (Hofmann 1976; Shukovsky & Halfen 1976; Horodyski *et al.* 1977;

Golubic & Barghoorn 1977; Haxo *et al.* 1987), such features are not considered significant in considering the genus-level taxonomy of fossil forms; in a natural taxonomy they may even find limited application to species differentiation. Thus, a number of previously named genera fall clearly within the morphological limits of *Oscillatoriopsis*, although, on the basis of unique subsidiary features, some of these may stand as distinct species within the genus (e.g., a new combination, *O. nodosa*, could accommodate the oscillatoriopsan *Halythrix nodosa* Schopf, 1968). The isodiametric cell dimensions of *O. awramikii* Wang *et al.*, 1983, and the larger *O. cuboides* Knoll *et al.*, 1988, are also sufficiently distinctive to warrant their retention as separate species. Most other 'species' of *Oscillatoriopsis* have cells considerably wider than long.

More than 75 species of *Oscillatoriopsis* in the nominal sense or as herein emended appear in the literature; 54 are accepted as belonging to the genus. Much of this purported diversity derives from relatively few and often very localized assemblages (Schopf 1968; Schopf & Blacic 1971; Knoll 1981; Zhang P. 1981, 1982), with almost all 'species' known from unique or very few specimens. The recognition of substantial intraspecific and taphonomic variation now dictates a more conservative taxonomy. The delineation of *Oscillatoriopsis* species should be based primarily on the identification of discrete size classes as determined from large populations (with the widest point of a trichome taken as the closest approximation of its original width). Most of the 113 specimens of *Oscillatoriopsis* recorded from the Svanbergfjellet Formation are circumscribed by a unimodal size-frequency distribution between 3 and 8 μm in diameter (= *O. obtusa*). The outliers from this distribution differ also in subsidiary qualitative characters, thus supporting their assignment to two other size classes (form species) that appear in the literature, *O. amadeus* n.comb. and *O. longa*. In summary, we recognize four basic species of *Oscillatoriopsis* with diameters less than 25 μm: *O. vermiformis* n.comb., 1–3 μm wide; *O. obtusa*, 3–8 μm wide; *O. amadeus* n.comb., 8–14 μm wide; and *O. longa*, 14–25 μm wide.

Oscillatoriopsis vermiformis (Schopf, 1968) Butterfield, n.comb.

Synonymy. – □1968 *Contortothrix vermiformis* Schopf, n.sp. – Schopf, p. 671, Pl. 79:7–8. □1968 *Cephalophytarion minutum* Schopf, n.sp. – Schopf, pp. 669–670, Pl. 78:9–12. □1968 *Anabaenidium Johnsonii* Schopf, n.sp. – Schopf, pp. 680–681, Pl. 81:4. □1968 *Archaeonema longicellularis* Schopf, n.sp. – Schopf, pp. 678–679, Pl. 80:11; *non* 1969 *Archaeonema longicellularis* Schopf, 1968 – Schopf & Barghoorn, p. 117, Pls. 21:2–4, 22:2–4.

Holotype. – Thin Section Bit/Spr 6–7, Paleobot. Coll. Harvard Univ. No. 58467; stage coordinates 19.4×107.4. (Bitter Springs Formation.) Schopf 1968, Pl. 79:7a–b.

Diagnosis. – A species of *Oscillatoriopsis* with trichomes 1–3 µm in diameter.

Discussion. – Some Svanbergfjellet *Oscillatoriopsis* measure as little as 3 µm in width, but these clearly represent the lower size limit of *O. obtusa*. By contrast, the small oscillatoriopsans that occur in the Ross River locality of the Bitter Springs Formation (Schopf 1968) appear to represent a distinct size class and thus warrant separate species status. All of the above synonymized taxa are accommodated by this new delineation: reexamination of the type specimen of *Archaeonema* shows it to be a small oscillatoriopsan, while the coiled habit of *Anabaenidium* and *Contortothrix* must be regarded as taxonomically unreliable in light of their rare occurrence and likely intraspecific variation. Overall, the most appropriate type species and specimen might have been taken from *Cephalophytarion minutum*; however, this species name has been previously occupied (i.e. *O. minuta* Lu & Zhu, 1985). *Oscillatoriopsis vermiformis* n.comb. has not been observed in the Svanbergfjellet assemblage.

Oscillatoriopsis obtusa Schopf, 1968, emend.

Fig. 24A–E, K

Synonymy. – □1968 *Oscillatoriopsis obtusa* Schopf, n. sp. – Schopf, p. 667, Pl. 77:8. □1968 *Caudiculophycus rivularioides* Schopf, n.sp. – Schopf, pp. 679–680, Pl. 79:3–6. □1968 *Cephalophytarion grande* Schopf, n.sp. – Schopf, p. 669, Pl. 78:1–4. □1971 *Caudiculophycus acuminatus*, n.sp. – Schopf & Blacic, p. 951, Pl. 105:7. □1971 *Cephalophytarion constrictum*, n.sp. – Schopf & Blacic, pp. 943–944, Pl. 105:1, 9. □1971 *Cephalophytarion delicatulum*, n.sp. – Schopf & Blacic, p. 946, Pl. 108:7. □1971 *Cephalophytarion grande* Schopf, 1968 – Schopf & Blacic, p. 956, Pl. 106:1, 2, 4, 9, 10. □1971 *Cephalophytarion laticellulosum*, n.sp. – Schopf & Blacic p. 944, Pls. 105:2, 6; 106:3. □1971 *Cephalophytarion variabile*, n.sp. – Schopf & Blacic, pp. 944–946, Pls. 107:2, 3, 5, 8; 108:3. □1971 *Oscillatoriopsis breviconvexa*, n.sp. – Schopf & Blacic, p. 943, Pl. 105:5. □1971 *Palaeolyngbya minor*, n.sp. – Schopf & Blacic, pp. 942–943, Pl. 105:4. □1971 *Partitiofilum gongyloides*, n.sp. – Schopf & Blacic, p. 947, Pls. 105:3; 106:6. □1977 *Oscillatoriopsis schopfii* sp. nov. – Oehler, p. 344, Fig. 13A. □1977 *Oscillatoriopsis psilata* sp. nov. – Maithy & Shukla, p. 179, Pl. 2:12. □1979 *Oscillatoriopsis anshanensis* Yin (sp. nov.) – Yin, p. 47, Pl. 2:10, 12. □1981 *Cephalophytarion taenia* n.sp. – Zhang Y., p. 493, Pl. 1:8–10. □1981 *Oscillatoriopsis jixianensis* Zhang sp. nov. – Zhang P., pp. 254–255, Pl. 1:1, 2, 4. □1981 *Oscillatoriopsis luozhuangensis* sp. nov. – Zhang P., p. 255. Pl. 1:5, 8. □1981 *Oscillatoriopsis qingshanensis* sp. nov. – Zhang P., p. 256, Pl. 1:3, 6. □1984 *Cephalophytarion turukhanicum* A. Weiss, sp. nov. – Weiss, pp. 104–105, Pl. 9:4. □1984 *Filiconstrictosus eniseicum* A. Weiss sp. nov. – Weiss, p. 105, Pl. 9:5. □1986 *Cephalophytarion pili-*

formis Mikhailova sp. nov. – Mikhailova, p. 35, Figs. 8–9. □1986 *Oscillatoriopsis parvula* Lie et Liu sp. nov. – Liu & Li, pp. 265–266, 270, Pl. 1:4. □1986 *Primorivularia dissimilara* Hermann sp. nov. – Hermann, p. 39, Figs. 13–14. □1986 *Primorivularia absoluta* Hermann sp. nov. – Hermann, pp. 39–40, Fig. 12. □1987 *Cyanonema disjuncta* Ogurtsova et Sergeev, sp. nov. – Ogurtsova & Sergeev, pp. 111–112, Pl. 9:3–4. □1987 *Obconicophycus minor* sp nov. – Yin, p. 479, Pl. 28:5, 8.

Holotype. – Thin Section Bit/Spr 6–7, Paleobot. Coll. Harvard Univ. No. 58448; stage coordinates 19.6×107.3. (Bitter Springs Formation.) Schopf 1968, Pl. 77:8.

Material. – One hundred nine (109) trichomes from shale samples 86-G-62, 86-G-28 and P-2945.

Emended diagnosis. – A species of *Oscillatoriopsis* with trichomes 3–8 µm in diameter; cells wider than long (length:width <1).

Description. – Cellular trichomes, 3–8 µm wide ($\bar{x}$ = 5.8 µm; s.d. = 1.2 µm; n = 109), with no sheath. Complete trichomes 50–200 µm long; usually occurring as isolated individuals, but in one instance as a loose association of ca. 50 trichomes extending over ca. 10 mm² of a shale bedding plane. Terminal cells of trichomes blunt, rounded, or tapered; typically one end rounded and the other gradually tapering. Cells 1.5–3.5 µm long; rarely constricted at septa.

Discussion. – Svanbergfjellet *O. obtusa* exhibit considerable variation in terminal cell morphology and overall length, but their otherwise uniform size (width) and habit argue convincingly against these features being applied to species-level taxonomy. Likewise, cell length seems to be somewhat variable, differing by up to 1 µm within a trichome and up to 2 µm between trichomes.

All of the above synonymized species fall within the size limits defined by the large, unimodal population of Svanbergfjellet *O. obtusa*. Given the disparate taphonomic and geologic histories that these various 'species' represent, their morphological variation does not exceed what might be reasonably expected from a single *Oscillatoria*-like precursor. On the other hand, the fundamental simplicity of such fossil trichomes precludes a natural taxonomy, and *O. obtusa* may well represent a diverse, though now undifferentiable assortment of taxa and physiologies (p. 13).

Oscillatoriopsis amadeus (Schopf & Blacic, 1971) Butterfield, n.comb.

Fig. 23E

Synonymy. – □1971 *Obconicophycus amadeus*, n.sp. – Schopf & Blacic, p. 950, Pl. 107:1a–b. □1982 *Oscillatoriopsis media* n.sp. – Mendelson & Schopf, p. 64–65, Pl. 4:3, 5, 6. □1982 *Oscillatoriopsis acuta* sp. nov. – Zhang P., pp. 36, 40,

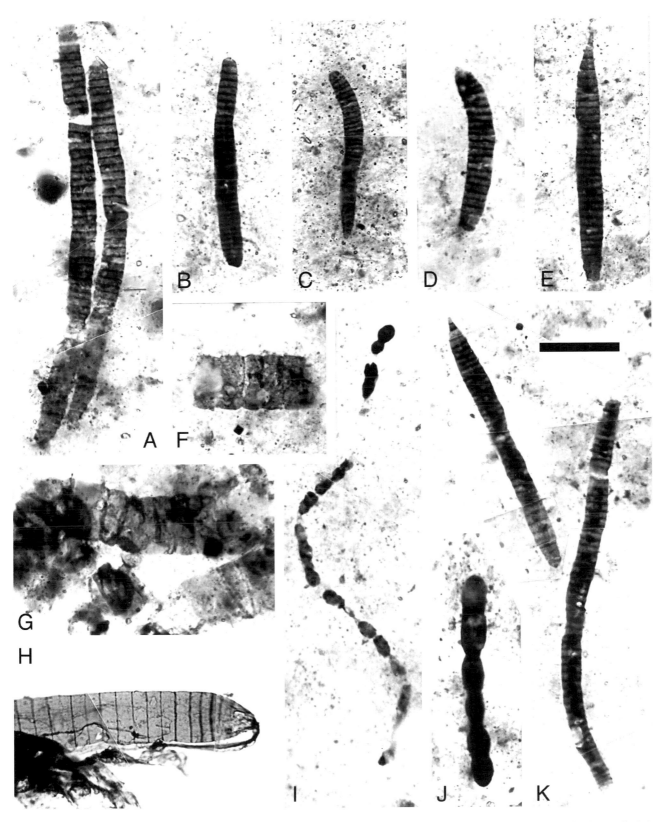

Fig. 24. Oscillatoriopsis obtusa Schopf (A–E, K); *Oscillatoriopsis longa* Timofeev & Hermann (F, G); *Tortunema Wernadskii* (Schepeleva) n.comb. (H); *Veteronostocale amoenum* Schopf & Blacic. (I); *Cyanonema* sp. (J). From shales of the Algal Dolomite Member, Geerabukta. Scale bar in K equals 25 μm. □A. HUPC 62869; 86-G-62-11 (N-27-2). □B. 86-G-62 (destroyed). □C. HUPC 62870; 86-G-62-1 (R-38-2). □D. HUPC 62871; 86-G-62-40 (L-46-4). □E. HUPC 62872; 86-G-62-65 (R-30-2). □F. HUPC 62873; 86-G-61-7 (Q-45-0). □G. HUPC 62874; 86-G-61-7 (Q-45-0). □H. HUPC 62875; 86-G-62-41M (M-35-1); detail of Fig. 27C; note that the pseudoseptate sheath collapsed as a unit rather than as individual cells. □I. HUPC 62876; 86-G-62-6 (G-39-2). □J. HUPC 62877; 86-G-62-10 (P-22-1). □K. HUPC 62878; 86-G-62-43 (P-12-2); with cytoplasmic residues preserved within individual cells.

Pl. 1:4a–b. ☐1982 *Oscillatoriopsis doliocellularis* sp. nov. – Zhang P., pp. 37, 40, Pl. 1:6. ☐1982 *O. formosa* sp. nov. – Zhang P., pp. 37, 40, Pl. 1:9. ☐1983 *Oscillatoriopsis taimirica* Schenfil, sp. nov. – Schenfil, p. 473, Pl. 1:2. ☐1987 *Oscillatoriopsis connectens* Zhang, sp. nov.(MS) – Zhang, p. 268, Pl. 1:5–6. ☐1989 *Cephalophytarion majesticum* Allison n.sp. – Allison & Awramik, pp. 273–274, Figs. 8.7–8.8.

Holotype. – Thin Section TBS-22-4-G; stage coordinates 50.5/115.0; Paleobot. Coll. Harvard Univ. 58579. (Bitter Springs Formation.) Schopf & Blacic 1971, Pl. 107:1a–b.

Material. – One (1) specimen in shale sample 86-G-62.

Diagnosis. – A species of *Oscillatoriopsis* with trichomes 8–14 µm in diameter; cells wider than long (length:width <1).

Description. – Cellular trichome, 10 µm in diameter, with no sheath; no terminal tapering. Cells ca. 4 µm long and narrowly separated from adjacent cells.

Discussion. – A single specimen of *Oscillatoriopsis* in the Algal Dolomite Member falls between the size distributions of *O. obtusa* and the larger *O. longa* (see below). Given its additional qualitative differences (all cells separated and lack of terminal taper) it is reasonably assigned to a separate taxon, *O. amadeus* n.comb. (alternatively, this specimen may have derived from a pseudocellular sheath such as *Cephalonyx* (cf. Fig. 23A).

The two original Bitter Springs specimens of *Oscillatoriopsis amadeus* n.comb. (= *Obconicophycus amadeus*) do not provide a useful size range for the species. On purely artificial grounds the size range falling between that of *O. obtusa* and *O. longa* might be suggested as it offers approximately the same degree of variation in trichome width as that documented for *O. obtusa*. However, as the delineation of a natural taxonomic group, this 8–14 µm size range awaits confirmation from a single large population.

Oscillatoriopsis longa Timofeev & Hermann, 1979, emend.

Fig. 24F–G

Synonymy. – ☐1979 *Oscillatoriopsis longum* Timofeev et Hermann, sp. nov. – Timofeev & Hermann, p. 139, Pl. 29:3, 4. ☐1982 *Oscillatoriopsis major* Liu sp. nov. – Liu, p. 148, Pl. 11:1. ☐1982 *Hyalothecopsis nanshanensis* gen. et sp. nov. – Zhang P., pp. 35–36, 39, Pl. 1:2. ☐1982 *Hyalothecopsis sinica* gen. et sp. nov. – Zhang P., pp. 35, 39, Pl. 1:1. ☐1983 *Halythrix leningradica* Schenf. sp. nov. – Schenfil, p. 473, Pl. 1:1. ☐1983 *Oscillatoriopsis variabilis* sp. nov. – Strother *et al.*, p. 26, Pl. 3:3–6, 11. ☐1984 *Oscillatoriopsis princeps* sp. nov. – Zhang & Yan, pp. 198, 203, Pl. 1:6. ☐1984 *Oscillatoriopsis aculeata* sp. nov. – Zhang & Yan, pp. 199, 203, Pl. 1:7. ☐1986 *Oscillatoriopsis connectens* sp. nov. – Zhang & Gu, pp. 323–324, 329, Pl. 1:10. ☐1986 *Oscillatoriopsis strictura* sp. nov. –

Zhang & Gu, pp. 324, 329–330, Pl. 1:8. ☐1986 *Oscillatoriopsis valida* sp. nov. – Zhang & Gu, pp. 324, 330, Pl. 1:9. ☐1986 *Oscillatoriopsis planaria* sp. nov. – Zhang & Gu, pp. 324, 330, Pl. 1:12. ☐1989 *Partitiofilum tungusum* Mikhailova, sp. nov. – Jankauskas *et al.*, p. 118, Pl. 27:4. ☐1991 *Filiconstrictosus magnus* Yakschin, sp. nov. – Yakschin, pp. 32–33, Pl. 11:2.

Holotype. – Slide No 19/6–76/6. Upper Precambrian, Lakhanda Formation, R. Maia, Khabarovsk District. Timofeev & Hermann 1979, Pl. 29:3.

Material. – Three (3) trichome fragments from shale samples 86-G-61 and 86-G-28.

Emended diagnosis. – A species of *Oscillatoriopsis* with trichomes 14–25 µm in diameter.

Discussion. – As with *O. amadeus*, a natural accounting of *O. longa* suffers from inadequate population sizes and/or detailed documentation. The upper size limit of 25 µm is taken from that originally suggested for *O. longa* and *Partitiofilum tungusum*; the lower is suggested by the dearth of reported trichomes 13–14 µm wide, and the absence of trichomes >14 µm wide in most assemblages containing smaller oscillatoriopsans. That the type specimen of *O. longa* is from shale rather than three-dimensionally preserved in chert is not considered significant as dimensions are expected to be the same in either context (p. 12). Instances of cellular disaggregation show the Svanbergfjellet specimens to be true trichomes rather than pseudoseptate sheaths.

Genus *Palaeolyngbya* Schopf, 1968, emend.

Synonymy. – ☐1968 *Palaeolyngbya* Schopf, n.gen. – Schopf, p. 665. ☐1974 *Paleolyngbya* Schopf, 1968 – Hermann, pp. 8–9. ☐1976 *Rhicnonema* n.gen. – Hofmann, p. 1053. ☐*non* 1971 *Palaeolyngbya minor*, n.sp. – Schopf & Blacic, pp. 942–943. Pl. 105:4 (= *Oscillatoriopsis*). ☐*nec* 1978 *Palaeolyngbya sinica* Yin et Li (sp. nov.) – Yin & Li, p. 89, Pl. 7:10. ☐*nec* 1983 *Palaeolyngbya spiralis* n.sp. – Wang *et al.*, pp. 162–164, Fig. 22:1, 2, 3, 7, 9. (= *Obruchevella*). ☐*nec* 1982 *Palaeolyngbya candeda* Luo et Wang, sp. nov. – Luo *et al.*, p. 26, Pl. 1:3. ☐*nec* 1982 *Palaeolyngbya vorticellata* Luo et Wang, sp. nov. – Luo *et al.*, p. 26, Pl. 1:5, 11 (= *Obruchevella*). ☐*nec* 1982 *Palaeolyngbya vermiformis* Luo et Wang, sp. nov. – Luo *et al.*, p. 27, Pl. 1:6, 8, 10. ☐*nec* 1983 *Palaeolyngbya elliptica* Zhang, sp. nov. – Zhang P., pp. 214, 219, Pl. 2:6. ☐*nec* 1991 *Palaeolyngbia* [*sic*] *giganteus* Yakschin, sp. nov. – Yakschin, p. 33, Pl. 12:1–2.

Type species. – *Palaeolyngbya Barghoorniana* Schopf, 1968, pp. 665–666.

Emended diagnosis. – Unbranched, uniseriate, multicellular trichomes within a single unornamented filamentous envelope.

Discussion. – As a form taxon *Palaeolyngbya* will certainly intergrade with other filamentous microfossils (cf. Hofmann 1976); We suggest that the name be reserved for smooth-walled filamentous sheaths that contain a regular uniseriate array of prominently preserved cells. Thus, the occasional preservation of cells in *Siphonophycus* (Fig. 27D), *Tortunema* (Fig. 27A), or *Rugosoopsis* (Fig. 25A–B) do not warrant their inclusion in *Palaeolyngbya*. Similarly, the regular helices of *P. spiralis* Wang et al., 1983, along with its ambiguously preserved cells, argue for its transfer to *Obruchevella*.

Cells within filamentous sheaths often show signs of severe shrinkage and their dimensions cannot be considered as useful characters, even within a form taxonomy. Uncollapsed sheath diameter is therefore the principal criterion for determining species of *Palaeolyngbya*.

Palaeolyngbya catenata Hermann, 1974

Fig. 25F–G

Synonymy. – ☐1974 *Paleolyngbya* [*sic*] *catenata* sp. n. – Hermann, pp. 8–9, Pl. 6:5. ☐1980 *Oscillatoriopsis robusta* n.sp. – Horodyski & Donaldson, pp. 149–152, Fig. 13H. ☐1981 *Palaeolyngbya maxima* n.sp. – Zhang Y., p. 495, Pl. 2:4, 6, 7. ☐1981 *Doushantuonema peatii* sp. nov. – Zhang Z., p. 204, Pl. 1a–c. ☐1982 *Scalariphycus tianzimiaoensis* Song (gen.et sp. nov.) – Song, p. 218, Pl. 32:9–11.

Material. – Eight (8) specimens: 3 in shale sample 86-G-62; 5 in chert sample P-2628.

Diagnosis. – A species of *Palaeolyngbya* with sheaths 10–30 μm in diameter.

Description. – Filamentous sheaths, 8–16 μm wide ($\bar{x} = 12.1$ μm; s.d. = 2.5 μm; $n = 8$), with a uniseriate array of partially degraded cells.

Discussion. – Originally described from relatively few specimens, *P. catenata* may be subject to some future modification with regard to size range; the holotype measures ca. 15 μm across. Svanbergfjellet *P. catenata* occur both in shales (Fig. 25G) and in subtidal silicified carbonates (Fig. 25F).

Palaeolyngbya hebeiensis Zhang & Yan, 1984, emend.

Fig. 25E, H

Synonymy. – ☐1984 *Palaeolyngbya hebeiensis* sp. nov. – Zhang & Yan, pp. 199, 203, Pl. 1:8. ☐1985 *Palaeolyngbya crassa* sp. nov. – Luo, pp. 172–173, 176, Pl. 2:1–4, 4–7. ☐1986 *Palaeolyngbya conicus* [*sic*] Liu et Lie sp. nov. – Liu & Li, pp. 266, 270, Pl. 1:5. ☐1986 *Palaeolyngbya sphaerocephala* Hermann et Pylina sp. nov. – Hermann, p. 38, Figs. 2–9.

Holotype. – 82029-1. (Gaoyuzhuang Formation.) Zhang & Yan 1984, Pl. 1:8.

Material. – Four (4) specimens from shale sample 86-G-62.

Emended diagnosis. – A species of *Palaeolyngbya* with sheaths 30–60 μm in diameter.

Description. – Filamentous sheaths, 36–46 μm wide ($\bar{x} = 42.3$ μm; s.d. = 3.8 μm; $n = 4$), with a uniseriate array of shrunken or partially degraded cells.

Discussion. – *Palaeolyngbya hebeiensis* was originally described from a single indifferently preserved fragment from the Mesoproterozoic Gaoyuzhuang Formation, China. The diameter of its 'hyaline' sheath was not stated, but appears to be little more than that of the cells (ca. 35 μm). *Palaeolyngbya sphaerocephala* is of this same size class (34–42.5 μm) and provides a population-based lower size limit for the taxon; the upper limit is suggested by the *P. crassa* population.

Palaeolyngbya hebeiensis can clearly grade (taphonomically) into the form genera *Rugosoopsis* (Fig. 25E; note the single transverse wrap outside the sheath), *Tortunema* (e.g., Hermann 1986, Figs. 7–8), and, of course, *Siphonophycus*. The specimen in Fig. 25H is unique in that the trichome has remained fully intact rather than separating into the more typical series of isolated cells.

Palaeolyngbya spp.

Fig. 25I

Material. – Four (4) measured specimens from shale sample 86-G-62.

Discussion. – Narrow, transversely banded filaments are conspicuous in the Svanbergfjellet shales, generally co-occurring with a variety of sheathed filamentous taxa; in some instances they can be observed to be physically contiguous. Close examination shows them to be uniseriate series of shrunken cells about which an enveloping sheath has collapsed, much as would be expected if cells had been preserved within the collapsed portion of the sheath shown in Fig. 24H. As the dimension of shrunken cells cannot be taken as a reliable taxonomic character, these *Palaeolyngbya* fossils are not classifiable to the species level.

Genus *Rugosoopsis* Timofeev & Hermann, 1979, emend.

Synonymy. – ☐1979 *Rugosoopsis* Timofeev et Hermann.gen. nov. – Timofeev & Hermann, p. 139. ☐1980 *Plicatidium* Jankauskas, gen. nov. – Jankauskas 1980b, p. 109.

Type species. – *Rugosoopsis tenuis* Timofeev & Hermann, 1979, p. 139.

Emended diagnosis. – Bi-layered filamentous sheaths; inner sheath smooth or pseudoseptate; outer sheath with a prominent transverse fabric.

Description. – *Siphonophycus*-like or *Tortunema*-like filamentous sheaths, wholly or partially enclosed in a second layer having a pronounced transverse fabric. Atrophied cells occasionally preserved within the inner sheath.

Discussion. – Differential preservation of the outer sheath of *Rugosoopsis* results in a wide, but superficial, variation in form. When largely retained it imparts a finely plicated or (pseudo-)cellular texture to filaments (Fig. 25A–B). More commonly, and to varying degrees, this outer layer 'unravels', exposing portions of the underlying pseudoseptate or smooth-walled sheaths (Fig. 25C–E); the unraveled sheath often remains associated as an entangled 'filamentous' halo (Figs. 25I, 27A, B). A clear gradation of *Rugosoopsis* into *Tortunema* (Fig. 27B), *Siphonophycus* (Fig. 25A–D), and *Palaeolyngbya* (Fig. 25E: note the single outer 'wrap' at the arrow) suggests that these various forms may all belong to the same *natural* taxon. *Cephalonyx* also has transverse banding outside(?) the filamentous sheath, but these are related specifically to the prior position of cells; the banding in *Rugosoopsis* is independent of the trichome.

Multiple extracellular sheaths are characteristic of a number of modern filamentous cyanobacteria, most notably species of *Scytonema* and *Lyngbya*. As with fossil *Rugosoopsis*, these multiple sheaths may not all be of a similar construction. For example, in their examination of degrading *L. aestuarii* at Laguna Mormona, Horodyski *et al.* (1977) distinguished an inner pliant-walled sheath from a clearly more rigid outer layer (Horodyski *et al.* 1977, Fig. 6M; compare with Fig. 25C herein). Interestingly, the degrading outer layer of *L. aestuarii* appears to develop a substantial *Rugosoopsis*-like transverse fabric (Horodyski *et al.* 1977, Fig. 6Q). Most *Rugosoopsis* are thus reasonably interpreted as the extracellular sheaths of oscillatoriacean cyanobacteria.

Pjatiletov (1988) emended *Rugosoopsis* to include *Plicatidium* Jankauskas, 1980(b), and, on the basis of large populations, found a distinct break in sheath size-frequency distribution at ca. 50 μm diameter; those less than 50 μm were considered *R. tenuis*, those larger, *R. latus* (Jankauskas, 1980). He named an additional large form (60–150 μm diameter), *R. rugososiusculus*, based on purported differences in the transverse banding.

Rugosoopsis tenuis Timofeev & Hermann, 1979, emend.

Figs. 25A–D, 27B

Synonymy. – □1979 *Rugosoopsis tenuis* Timofeev et Hermann.gen. et sp. nov. – Timofeev & Hermann, p. 139, Pl. 29:5, 7. □1982 *Tubulosa corrugata* Assejeva gen. et sp. nov. – Assejeva, p. 13, Pl. 2:10, 11. □1984 *Karamia costata* Kolosov, sp. nov. – Kolosov, pp. 41–42, Pl. 6:1. □1984 *Karamia jazmirii* (Kolosov), 1982 – Kolosov, p. 40–41, Pl. 5:1, 2; *non* Pl. 4:2. □1984 *Karamia segmentata* Kolosov, sp. nov. – Kolosov,

p. 40, Pls. 3:1; 4:1. □1987 *Siphonophycus costatus* Jankauskas, 1980 – Yin, p. 480, Pl. 11:1–4.

Holotype. – Slide No 1-22/1-77/1. Upper Precambrian, Lakhanda Formation, R. Maia, Khabarovsk District. Timofeev & Hermann 1979, Pl. 29:7.

Material. – Fifty six (56) specimens: 54 from shale samples 86-G-62 and 86-G-61; 2 from chert samples P-2664-4A and 86-G-14-1A.

Emended diagnosis. – A species of *Rugosoopsis* with sheaths less than 60 μm in diameter.

Discussion. – Svanbergfjellet *Rugosoopsis* range from 7 μm to 57 μm in diameter with a mode at ca. 30 μm ($\bar{x}$ = 29.1 μm; s.d. = 13.4 μm; n = 56), but there is little indication that more than a single population is represented. This substantially increases the 30.0–37.5 μm range proposed in the original description of *R. tenuis* (Timofeev & Hermann 1979) and marginally exceeds that suggested by Pjatiletov (1988). Most *R. tenuis* have been reported from shale facies; however, two small (ca. 10 μm diameter) specimens were recorded from silicified carbonates in the Lower Dolomite and Algal Dolomite members; the silicified *Siphonophycus* specimen in Fig. 21E has associated material that may represent a disaggregated outer sheath.

Genus *Siphonophycus* Schopf, 1968, emend. Knoll, Swett & Mark, 1991

Synonymy. – □1968 *Siphonophycus* Schopf, n.gen. – Schopf, p. 671. □*non* 1989 *Siphonophycus attenuatum* A. Weiss, sp. nov. – Jankauskas *et al.*, pp. 121–122, Pl. 25:6–7 (= *Cephalonyx*).

Type species. – *Siphonophycus kestron* Schopf, 1968, p. 671.

Discussion. – Following a number of taxonomic revisions (e.g., Zhang Z. 1982, Pjatiletov 1988; Knoll *et al.* 1991), all unbranched, smooth-walled, (originally) tubular sheaths have come to be classified under the form genus *Siphonophycus*. (*Taeniatum* Sin, 1962, or *Gunflintia* Barghoorn, 1965, are likely senior synonyms, but these names have not been broadly employed.) Species are in turn defined on the basis of discrete size (width) classes as determined from reasonably large populations (Knoll *et al.* 1991). Although correct in principle, these above revisions were typically carried out at the genus level, thereby failing to recognize the taxonomic priority of some of the synonymized species (e.g., *S. typicum* n.comb. and *S. solidum* n.comb.).

Siphonophycus is clearly an artificial form taxon that will inevitably include a disparate range of organisms – bacterial, cyanobacterial, protistan and fungal. Even so, among Proterozoic filaments there is a broad recurring modality (Knoll 1982) that suggests some underlying natural distinctions. At

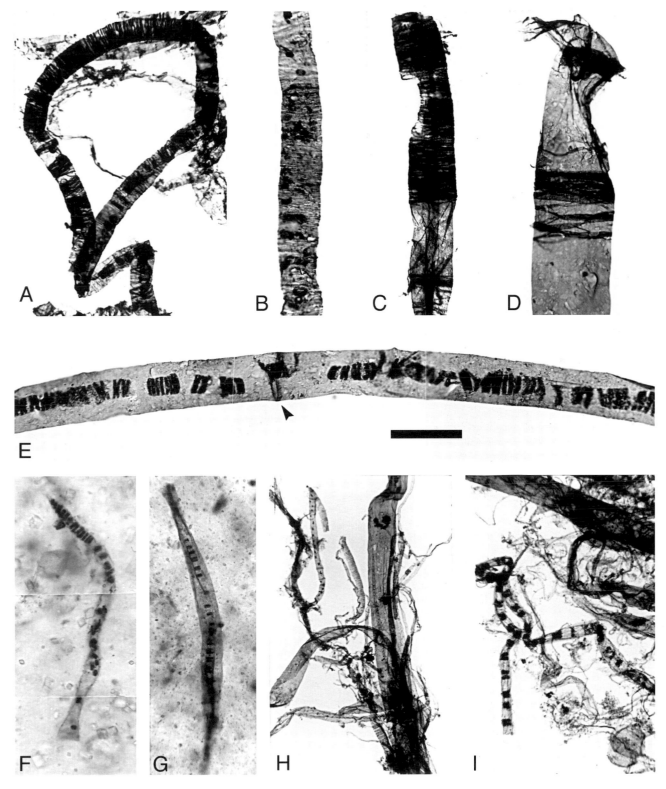

Fig. 25. Rugosoopsis tenuis Timofeev & Hermann (A–D); *Palaeolyngbya hebeiensis* Zhang & Yan (E); *Palaeolyngbya catenata* Hermann (F, G); *Palaeolyngbya/Rugosoopsis/Siphonophycus* (H, I). From shales of the Algal Dolomite Member, Geerabukta (A–E, G–I), and cherts of the Lower Dolomite Member, Polarisbreen (F). Scale bar in E equals 50 μm for A, B, D–G, I; 80 μm for C, H. □A. HUPC 62879; 86-G-62-126M (J-36-1); with cell remnants within the sheath. □B. HUPC 62880; 86-G-62-126M (K-40-3); with cell remnants. □C. HUPC 62881; 86-G-62-148M (P-43-3); outer sheath partially removed. □D. HUPC 62882; 86-G-62-120M (J-28-4); outer sheath largely removed. □E. HUPC 62883; 86-G-62-132M (O-32-2); with single transverse 'wrap' at the arrow. □F. HUPC 62884; P-2628-1A (D-58-1); in chert. □G. HUPC 62885; 86-G-62-67 (P-11-4); in shale. □H. HUPC 62886; 86-G-62-93M (P-46-0); various filamentous form taxa that may represent a single biological entity. Note the narrow cellular trichome within the otherwise *Siphonophycus solidum* (Golub) n.comb. filament. □I. HUPC 62887; 86-G-62-115M (M-29-4); series of atrophied cells within a collapsed sheath.

the very least it provides a useful classification scheme that, as it happens, delineates species on the basis of geometrically increasing size (diameter) ranges. Thus: *S. septatum* – 1–2 μm; *S. robustum* – 2–4 μm; *S. typicum* n.comb. – 4–8 μm; *S. kestron* – 8–16 μm; and *S. solidum* n.comb. – 16–32 μm. The Svanbergfjellet assemblage includes examples of all of these, as well as of a smaller (ca. 0.5 μm in diameter) form, *S. thulenema* n.sp.

Siphonophycus thulenema Butterfield, n.sp.

Fig. 22I

Holotype. – HUPC 62718, Fig. 22I; Slide 86-G-62-46, England-Finder coordinates R-28-3.

Type locality. – Algal Dolomite Member, Svanbergfjellet Formation, Geerabukta (79°35'30"N, 17°44"E); 55 m above base of member.

Material. – Three (3) populations from shale sample 86-G-62.

Diagnosis. – A species of *Siphonophycus* ca. 0.5 μm in diameter.

Description. – Unbranched, smooth-walled, filamentous microfossils, 0.5 μm in diameter and up to several hundred micrometers long. Typically gregarious, forming both small sub-parallel associations and relatively extensive mats.

Discussion. – *Archaeotrichion contortum* has conventionally been the taxon to which submicrometer-diameter fossil filaments were assigned; however, reexamination of the type specimen shows it to fall within the bounds of *Siphonophycus septatum* (see below). *Siphonophycus thulenema* n.sp. represents a smaller size class quite distinct from *S. septatum* or any other previously described taxa, and its repeated and well-defined occurrences in bedding-parallel thin-section show it not to be a degradational variant of some larger form; its width does not vary appreciably from 0.5 μm. As with other small species of *Siphonophycus* (Knoll *et al.* 1991), physiologic and higher taxonomic determination of *S. thulenema* remains speculative.

Etymology. – From the Greek *thoule* – northernmost, and *nema* – thread.

Siphonophycus septatum (Schopf, 1968) Knoll *et al.*, 1991

Figs. 10H, 22G–H

Synonymy. – □1968 *Tenuofilum septatum* Schopf, n.sp. – Schopf, p. 679, Pl. 86:10–12. □1968 *Archaeotrichion contortum* Schopf, n.sp. – Schopf, p. 686, Pl. 86:1–2. □1980

Eomycetopsis? campylomitus n.sp. – Lo, pp. 143–144, Pl. 1:9–11. □1982 *Allachjunica tenuiuscula* Kolosov, gen. et sp. nov. – Kolosov, p. 80, Pl. 14:1. □1982 *Judomophyton microscopicum* Kolosov, gen. et sp. nov. – Kolosov, p. 75, Pl. 11:1. □1982 *Siphonophycus chuii* Liu sp. nov. – Liu, p. 148, Pl. 11:5, 8, 11. □1984 *Eophormidium capitatum* Xu sp. nov. – Xu, pp. 219, 314, Pl. 1:5. □1989 *Archaeotrichion lacerum* Hermann, sp. nov. – Jankauskas *et al.*, p. 88, Pl. 39:4a–b. □1991 *Siphonophycus septatum* comb. nov. – Knoll *et al.*, p. 565, Fig. 10.2.

Material. – Two relatively extensive 'mat' populations and several isolated specimens, from shale sample 86-G-62; 15 occurrences within *Valkyria* n.gen.

Description. – Unbranched, nonseptate, smooth-walled filamentous microfossils, 1–2 μm in diameter.

Discussion. – *Siphonophycus septatum* in the Svanbergfjellet Formation is found only in shale facies, where it occurs as isolated individuals (Fig. 22H); in loose, sometimes regularly criss-crossing associations; or, in one instance, as a tightly intertwined triple helix (Fig. 22G; cf. *Flagellis* Assejeva, 1982). *Siphonophycus septatum* filaments are also repeatedly found lining a central space in the multicellular problematicum *Valkyria* n.gen., suggesting their possible role as biodegraders. Alternatively, they might be interpreted as symbiotic with (or parasitic upon) the larger organism. In any event, this variety of habits clearly illustrates the incorporation of diverse organisms within the form taxon *S. septatum*.

Reexamination of the type specimen of *Archaeotrichion contortum* shows it, where not collapsed, to be a full micrometer in diameter and therefore a junior synonym of *S. septatum*. The type material of *A. lacerum* likewise has an original diameter of at least 1 μm, while its diagnosed intertwining habit does not appear to warrant separate species designation.

Siphonophycus robustum (Schopf, 1968) Knoll *et al.*, 1991

Fig. 26A, G

Synonymy. – □1968 *Eomycetopsis robusta* Schopf, n.sp. – Schopf, p. 685, Pls. 82:2–3; 83:1–4. □1968 *Eomycetopsis filiformis* Schopf, n.sp. – Schopf, pp. 685–686, Pls. 82:1, 4; 83:5–8. □1969 *Archaeonema longicellularis* Schopf, 1968 –

Fig. 26. *Siphonophycus robustum* (Schopf) Knoll (A, G); *Siphonophycus typicum* (Hermann) n.comb. (B, H, I); Filaments mineralized by goethite(?) (C–F). From shales of the Algal Dolomite Member, Geerabukta (A–G), and cherts of the Lower Dolomite Member, Geerabukta (H, I). Scale bar in H equals 80 μm for A, B, E, G–I; 200 μm for C; 50 μm for D, F. □A. HUPC 62888; 86-G-62-60 (N-29-3); bedding-parallel thin-section, shale. □B. HUPC 62889; 86-G-30-6 (N-16-2); bedding-parallel thin section, shale. □C. HUPC 62890; 86-G-30-4BP; bedding plane specimen. □D. HUPC 62891; 86-G-30-3 (O-37-1); bedding-parallel thin section showing

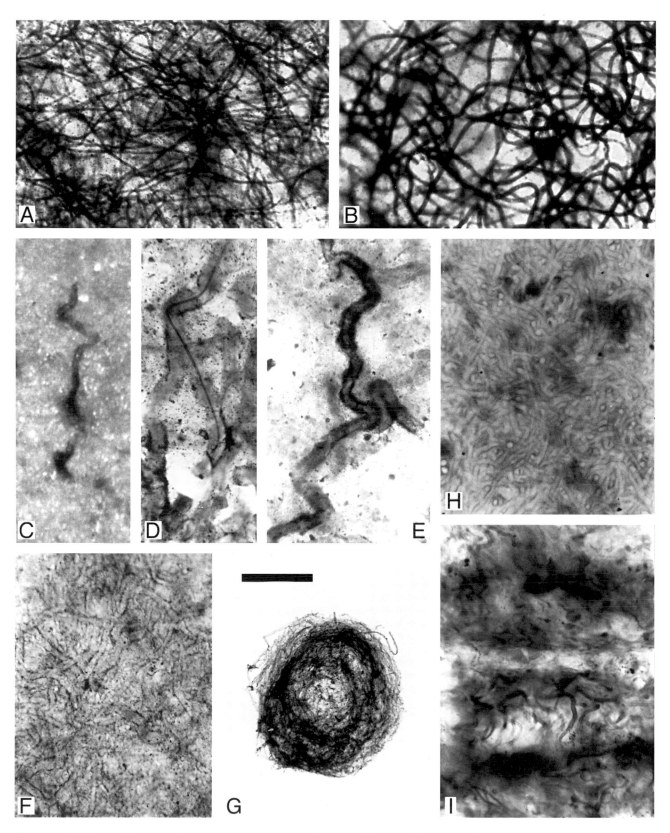

filaments with thick mineral overgrowths. □E. HUPC 62892; 86-G-30-5 (G-38-3); bedding-parallel thin section showing diagenetically induced 'branching'. □F. 86-G-30 (destroyed); bedding-parallel thin section showing entangled, mat-like habit. □G. HUPC 62893; 86-G-62-173M (K-34-2); small *Nostoc*-ball-like aggregation of filaments. □H. HUPC 62894; 86-G-14-1A (R-65-4); silicified intraclast composed of dense, unoriented filamentous mat; no allochthonous sedimentary particles. □I. HUPC 62895; 86-G-15-1 (T-46-3); silicified intraclast composed of alternating vertically and horizontally oriented filamentous mat; substantial input of allochthonous sediments.

Schopf & Barghoorn, p. 117, Pls. 21:1–4; 22:2–4. □1977 *Eomycetopsis psilata* sp. nov. – Maithy & Shukla, p. 180, Pl. 2:15. □1978 *Beckspringia communis* n.sp. – Licari, pp. 779–780. Pl. 1:3–7. □1982 *Acranella granulata* Kolosov, gen et sp. nov. – Kolosov, p. 81, Pl. 13:1a–b. □1982 *Allachjunica daedalea* Kolosov, gen et sp. nov. – Kolosov, pp. 79–80, Pl. 12:2a–b. □1982 *Judomophyton minisculum* Kolosov, gen. et sp. nov. – Kolosov, p. 76, Pl. 11:2. □1983 *Eomycetopsis polesicus* Assejeva sp. nov. – Assejeva, p. 170, Pl. 7:12. □1984 *Eophormidium liangii* Xu sp. nov. – Xu, pp. 219, 314, Pl. 3:3–4. □1984 *Eophormidium semicirculare* Xu sp. nov. – Xu, pp. 219, 314–315, Pl. 2:1–2. □1984 *Oscillatoriopsis acuminata* Xu sp. nov. – Xu, pp. 218, 312, Pl. 1:3, 4, 6. □1984 *Oscillatoriopsis disciformis* Xu sp. nov. – Xu, pp. 218–219, 313, Pl. 3:7. □1984 *Oscillatoriopsis glabra* Xu sp. nov. – Xu, pp. 219, 313, Pls. 2:6, 8A; 3:9, 11. □1984 *Oscillatoriopsis hemisphaerica* Xu sp. nov. – Xu, pp. 218, 312, Pls. 1:7–8; 2:12. □1984 *Oscillatoriopsis tuberculata* Xu sp. nov. – Xu, pp. 219, 313, Pl. 1:1–2. □1984 *Schizothropsis caudata* Xu sp. nov. – Xu, pp. 219, 315, Pls. 3:1–2, 10; 2:3–4. □1991 *Siphonophycus robustum* comb. nov. – Knoll *et al.*, p. 565, Fig. 10:3, 5.

Material. – Twenty-seven (27) mat-forming populations extending laterally for more than 1 mm: 22 in bedding-parallel thin section; 5 isolated by acid maceration. From shale samples 86-G-62, 86-G-61 and 86-G-30. Three (3) spheroidal 'colonies' from shale sample 86-G-62.

Description. – Unbranched, nonseptate, smooth-walled filamentous microfossils, 2–3 µm in diameter.

Discussion. – *Siphonophycus robustum* commonly serves as an accessory builder in silicified Proterozoic microbial-mat communities (e.g., Schopf 1968; Green *et al.* 1989; Knoll *et al.* 1991). It seems not to occur in Svanbergfjellet cherts; however, it is the principal constituent of shale-facies microbial mats, the entangled filaments commonly extending for nearly a centimeter in both bedding-parallel thin sections (Fig. 26A) and macerated material. *Siphonophycus robustum* also occurs as the sole constituent of circular (originally spheroidal?) structures up to several hundred micrometers in diameter (Fig. 26G) that may be broadly comparable to modern *Nostoc* balls. The ca. 4 mm spheroidal compression in Fig. 8F appears to have a similar filamentous construction, illustrating the potential confusion of such compound structures with large acritarchs (cf. Sun 1987a).

Siphonophycus typicum (Hermann, 1974) Butterfield, n.comb.

Figs. 23B–D, 26B, H, I

Synonymy. – □1974 *Leiothrichoides tipicus* gen. et sp.n. – Hermann, p. 7, Pl. 6:1–2. □1975 *Eomycetopsis cylindrica* sp. nov. – Maithy, p. 140, Pl. 4:27–28. □1975 *Eomycetopsis rugosa* sp. nov. – p. 140, Pl. 4:25–26. □1977 *Eomycetopsis pflugii*

sp. nov. – Maithy & Shukla, p. 180, Pl. 2:16. □1980 *Siphonophycus crassiusculum* n.sp. – Horodyski, p. 656, Pl. 1:6–7. □1980 *Eomycetopsis rimata* Jankauskas, sp. nov. – Jankauskas 1980b, pp. 111–112, Pl. 12:11. □1980 *Eomycetopsis? siberiensis* n.sp. – Lo, pp. 139–143, Pl. 1:1–8. □1981 *Siphonophycus inornatum* n.sp. – Zhang Y., pp. 491–493, Pl. 1:1, 3–5. □1982 *Judomophyton vulgatum* Kolosov, gen. et sp. nov. – Kolosov, p. 77, Pls. 11:4; 12:1a–b. □1982 *Judomophyton multum* Kolosov, gen. et sp. nov. – Kolosov, pp. 76–77, Pl. 11:3. □1982 *Sacharia crassa* Kolosov, gen. et sp. nov. – Kolosov, p. 79. Pl. 12:3. □1982 *Uraphyton rectum* Kolosov, gen. et sp. nov. – Kolosov, p. 82, Pl. 13:2. □1982 *Siphonophycus hughesii* sp. nov. – Nautiyal, p. 175, Fig. 1A–G. □1982 *Heliconema randomensis* sp. nov. – Nautiyal, p. 176, Fig. 1J–N. □1982 *Eomycetopsis pachysiphonia* Zhu, sp. nov. – Zhu, p. 6, Pl. 1:1–4. □1982 *Eomycetopsis crassiusculum* (Horodyski, 1980) comb. nov. – Zhang Z., pp. 455–456, Pl. 47:3–6, 9–13. □1985 *Eomycetopsis crassus* n.sp. – Yin 1985b, p. 180, Pls. 1:1–2, 5–6; 2:9. □non 1979 *Leiothrichoides typicus* Hermann – Timofeev & Hermann, p. 138, Pl. 29:1.

Holotype. – Slide No. 49/2T. Krasnoyarks District, Turukhansk Region, R. Miroedikha, Miroedikha Formation, Upper Riphean. Hermann 1974, Pl. 6:1-2.

Material. – Abundant populations in chert samples 86-G-8, 86-G-9, 86-G-14, 86-G-15, P-2664, 86-P-89 and P-3400. Eight (8) populations from shale samples 86-G-62 and 86-G-30.

Diagnosis. – A species of *Siphonophycus* with filament diameters 4–8 µm.

Description. – Unbranched, nonseptate, smooth-walled filamentous microfossils, 4–8 µm in diameter. Some specimens with discrete thickened or darkened intervals.

Discussion. – *Siphonophycus typicum* n.comb. is the dominant mat builder in stromatolitic facies of the Svanbergfjellet Formation. It occurs most commonly as extremely dense, sediment-free and unoriented microbial mat (Fig. 26H), sometimes in thicknesses of more than 1 cm. With the input of clastic material, a more oriented fabric of alternating vertical and horizontal filaments is developed (Fig. 26I); the higher taxonomic diversity of these sediment-bearing mats suggests deposition in less stressed, lower tidal-flat environments (Knoll *et al.* 1991). Both of these microbial-mat fabrics are recorded primarily from the intraclasts of silicified microbialite grainstones, and their paleoenvironmental interpretation is therefore inferential. The interstices of these grainstones, however, represent an additional, and *in situ*, paleoenvironment, which also preserves *S. typicum*. Here the filaments tend to be solitary, very straight, and commonly exhibit 10–100 µm long intervals of 'pointilistically' darkened sheath (Fig. 23B–D; p. 11; Sovetov & Schenfil 1977, Fig. 2e). *Siphonophycus typicum* is less regularly encountered in Svanbergfjellet shales but does occasionally occur as entangled mats (Fig. 26B). It is readily distinguished from co-occurring *S. robustum* by its darker and seemingly more rigid walls, as well as its larger size.

The above synonymy reflects the taxonomic adjustments necessitated by recent generic revisions of *Siphonophycus* (e.g., Pjatiletov 1988; Knoll *et al.* 1991). As *Leiothrichoides* is now a junior synonym of *Siphonophycus*, *S. typicum* n.comb. is the first named species defining a size range between that of *S. robustum* and *S. kestron* (the *S. typicum* holotype is ca. 6.5 µm wide). The paratype offered by Timofeev & Hermann (1979) somewhat exceeds the upper size limit of *S. typicum* (see Knoll 1982, Text-fig. 3 – *S. inornatum*) and is more appropriately assigned to *S. kestron*.

Siphonophycus kestron Schopf, 1968

Fig. 21D

Synonymy. – ☐1968 *Siphonophycus kestron* Schopf, n.sp. – Schopf, p. 671, Pl. 80:1–3. ☐1979 *Leiothrichoides typicus* Hermann – Timofeev & Hermann, pp. 138–139, Pl. 29:1. ☐1979 *Omalophyma angusta* Golub, sp. nov. – Golub, pp. 151–152, Pl. 30:13–18. ☐1979 *Isophyma stricta* Golub, sp. nov. – Golub, p. 154, Pl. 32:11–12. ☐1980 *Siphonophycus beltensis* n.sp. – Horodyski, pp. 654–656, Pl. 1:4. ☐1980 *Euryaulidion cylindratum* n.sp. – Lo, pp. 144–146, Pl. 2:1–3. ☐1980 *Siphonophycus indicus* sp. nov. – Nautiyal, p. 3, Fig. 1A. ☐1982 *Judomophyton unifarium* Kolosov, gen. et sp. nov. – Kolosov, p. 78, Pls. 11:5; 12:4. ☐1982 *Uraphyton distinctum* Kolosov, gen. et sp. nov. – Kolosov, pp. 81–82, Pl. 14:2a–b. ☐1982 *Uraphyton evolutum* Kolosov, gen. et sp. nov. – Kolosov, pp. 82–83, Pls. 14:3; 15:1. ☐1982 *Gunflintia brueckneri* sp. nov. – Nautiyal, pp. 175–176, Pl. 1H–I. ☐1984 *Siphonophycus laishuiensis* sp. nov. – Zhang & Yan, pp. 198, 203, Pl. 1:3. ☐1984 *Eomycetopsis contorta* Zhu, sp. nov. – Zhu *et al.*, pp. 173–174, 183, Pl. 3:1–3, 6. ☐1985 *Eomycetopsis lata* V. Golovenoc et M. Belova, sp. nov. – Golovenoc & Belova, p. 99, Pl. 7:4. ☐1985 *Siphonophycus ganjingziensis* (sp. nov.) – Bu, p. 210, Pl. 1:1–5. ☐1985 *Taeniatum punctosum* (sp. nov.) – Du, p. 162, Pl. 2:22–23. ☐1986 *Siphonophycus sinensis* sp. nov. – Zhang, pp. 32, 36, Pls. 1:1, 3; 2:4.

Material. – Rare specimens in chert sample P-2628 and shale sample 86-G-62.

Description. – Unbranched, nonseptate, smooth-walled filamentous microfossils, 8–16 µm in diameter.

Discussion. – *Siphonophycus* filaments with diameters between 8 and 16 µm are relatively rare in the Svanbergfjellet assemblage. Moreover, those that are recorded, both from chert and shale, are associated with filamentous material that may represent a disaggregated outer sheath comparable to that of *Rugosoopsis*.

Siphonophycus solidum (Golub, 1979) Butterfield, n.comb.

Figs. 25H–I, 27D

Synonymy. – ☐1979 *Omalophyma solida* Golub, sp. nov. – Golub, p. 151, Pl. 31:1–4, 7. ☐1979 *Omalophyma gracilis* Golub, sp. nov. – Golub, p. 151, Pl. 31:5–6, 8–9. ☐1979 *Solenophyma rudis*, sp. nov. – Golub, p. 153, Pl. 32:1–3. ☐1979 *Solenophyma tenuis* Golub, sp. nov. – Golub, p. 153, Pl. 32:4–6. ☐1980 *Leiothrichoides gracilis* Pjatiletov, sp. nov. – Pjatiletov, pp. 16–17, Pl. 4:4–5. ☐1982 *Uraphyton crassitunicatum* Kolosov, gen. et sp. nov. – Kolosov, p. 83, Pl. 15:3. ☐1982 *Uraphyton lenaicum* Kolosov, gen. et sp. nov. – Kolosov, p. 83, Pl. 15:2. ☐1984 *Siphonophycus capitaneum* n.sp. – Nyberg & Schopf, p. 753, Fig. 11E. ☐1987 *Siphonophycus transvaalensis* Beukes, Klein & Schopf, n. sp. – Klein *et al.*, p. 88, Fig. 5A–L. ☐1988 *Eomycetopsis grandis* Pjatiletov, sp. nov. – Pjatiletov, pp. 68–69, Pl. 7:2. ☐1991 *Siphonophycus capitaneum* Nyberg and Schopf, 1984 – Knoll *et al.*, p. 565, Fig. 3.4.

Holotype. – VSEGEI, Slide R-163/3, Rudnian collection; Upper Vendian, Smolensk Formation (upper part of the interval 747.8–767.3 m). Golub 1979, Pl. 31:1.

Material. – Twenty one (21) specimens from shale sample 86-G-62.

Diagnosis. – A species of *Siphonophycus* 16–32 µm in diameter.

Description. – Unbranched, nonseptate, smooth-walled filamentous microfossils, 16–32 µm in diameter ($\bar{x}$ = 22.9 µm; s.d. = 5.7 µm; n = 21).

Discussion. – A number of larger-diameter *Siphonophycus* species have appeared in the literature, but it is often not clear what, if any, biologically meaningful size distributions are being delineated. The above species of *Omalopyma* and *Solenophyma* range from 17–29 µm in diameter and specimens of *S. capitaneum* from 24–33 µm; *Leiothrichoides gracilis* is described as less than 26 µm and *Eomycetopsis grandis* as greater than 16 µm wide. In light of the size distribution of the Svanbergfjellet population, the size range 16–32 µm may circumscribe a possible natural grouping; it has the added, if entirely artificial, advantage of maintaining the delineation of *Siphonophycus* species by the geometric increase of their size (diameter) ranges (p. 64).

Genus *Tortunema* Hermann, 1976, emend.

Synonymy. – ☐1976 *Tortunema* Hermann gen.n. – Timofeev *et al.*, p. 39. ☐*non* 1976 *Tortunema eniseica* Hermann gen. et sp. n. – Timofeev *et al.*, p. 40, Pl. 12:4 (= *Siphonophycus*).

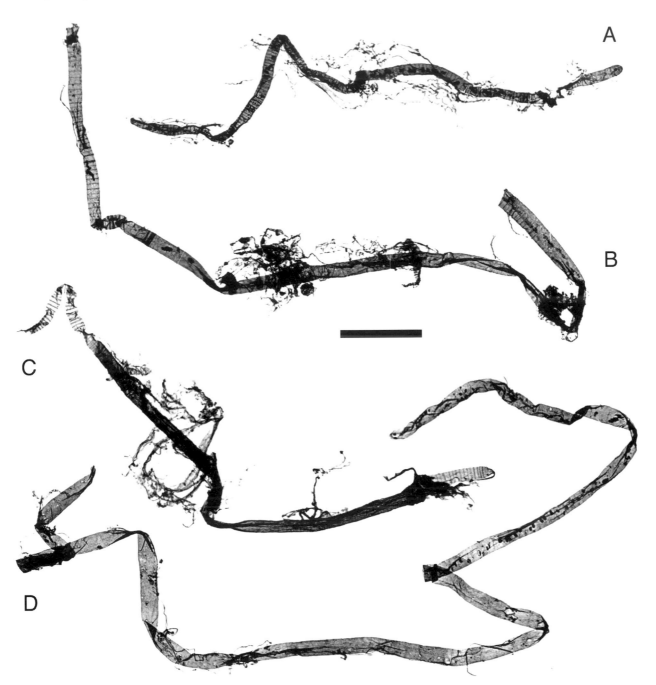

Fig. 27. Tortunema/Rugosoopsis/Siphonophycus. From shales of the Algal Dolomite Member, Geerabukta. Scale bar in B equals 100 μm. □A. HUPC 62896; 86-G-62-147M (R-21-2); pseudoseptate filament (*Tortunema*) containing freely transported cell residues and a halo of *Rugosoopsis*-like outer sheath material. □B. HUPC 62897; 86-G-62-233M (N-37-0); pseudoseptate filament (*Tortunema*) with localized sections that could be classified as *Siphonophycus* or *Rugosoopsis*, with halo of disaggregated outer sheath material. □C. HUPC 62875; 86-G-62-41M (M-35-1); pseudoseptate filament (*Tortunema*) with intact outer sheath; unlike the transverse fabric typical of *Rugosoopsis*, this outer layer appears to be oriented longitudinally; see Fig. 2H for detail of inner sheath. □D. HUPC 62898; 86-G-62-113M (L-30-0); *Siphonophycus soplidum* (Golub) n.comb., with dispersed cells.

Type species. – Tortunema Wernadskii (Schepeleva, 1960) n.comb., p. 170.

Emended diagnosis. – Unbranched filamentous sheaths with thin annular lines or thickenings (pseudoseptate filaments). Intervals between annulations usually less than sheath diameter and usually regular.

Discussion. – In contrast to truly cellular *Oscillatoriopsis*, *Tortunema* is simply a filamentous sheath upon which the position of trichome septa have been impressed (*vs.* the position of the cells themselves; see *Cephalonyx*); the distinction, however, may not always be clear. Pseudoseptate filaments can often be identified by the presence of cellular residues that have moved freely within the sheath (Fig. 27A;

Hermann 1986, Figs. 7–8); by patterns that grade longitudinally (and even laterally) into unambiguous *Siphonophycus*-type sheaths (Fig. 27B); by a tendency for differential 'septal' prominence or 'missing septa' (Fig. 24H); and/or by degradational collapse of the sheath rather than a disaggregation into separate cells (Figs. 24H, 27C). These criteria, however, may not decide all cases of apparently septate filaments; for example, had the *Tortunema* specimen in Fig. 27A not retained its few 'transported' cells, it would have been classified readily as *Oscillatoriopsis*. In light of the well-documented preservational bias of cyanobacterial sheaths over actual cell constituents (Golubic & Barghoorn 1977; Horodyski *et al.* 1977; Bauld 1981), pseudoseptate *Tortunema* is the preferred interpretation for otherwise ambiguous specimens. *Oscillatoriopsis*, *Cyanonema* and *Veteronostocale* are most usefully reserved for those filaments with *positive* evidence of true cellular preservation.

The earliest named genus to which pseudoseptate fossils might be ascribed is *Oscillatoriites* Zalessky, 1926; however, the large size (200 µm wide) and ambiguous illustration of the type species, *O. bertrandi* (Jurassic, Volga Basin), militate against its being applied to the much smaller Proterozoic occurrences. Thus Kolosov (1984) reformulated *Oscillatoriites Wernadskii* Schepeleva, 1960, as the type species of *Botuobia* Pjatiletov, 1979, with both he (Kolosov) and Jankauskas *et al.* (1989) distinguishing *Botuobia* from *Oscillatoriopsis* primarily on the basis of size. The two genera are indeed distinct, but not because of generically insignificant differences in dimension: *Botuobia Wernadskii*, indeed all the named species of *Botuobia*, are clearly extracellular sheaths bearing only the imprints of the cell septa they once contained. As *Tortunema sibirica* Hermann, 1976, is an earlier-named taxon of pseudoseptate filaments (similarly identified by its often 'missing septa'), *Botuobia* becomes its junior synonym, and the type species reverts to *T. Wernadskii* n.comb.

Lack of distinctive characters leading to the inclusion of disparate natural taxa, and differential preservation leading to an intergradation with other filamentous form taxa present obvious taxonomic problems for *Tortunema*. The latter point is exemplified by single filaments that could fall into as many as four different form genera depending on what portion is examined (Fig. 27B; p. 13). Even within the *Tortunema* form, a feature such as the spacing between annulations proves to be an unreliable taxonomic character (Pjatiletov 1979; Kolosov 1984). The only consistent approach to delineating species, then, appears to be on the basis of size (filament diameter). Under the present revision at least four species of *Tortunema* are recognized: *T. angusta* (Kolosov, 1984) n.comb. (sheaths less than 10 µm in diameter); *T. Wernadskii* (Schepeleva, 1960) n.comb. (sheaths 10–25 µm); *T. patomica* (Kolosov, 1982) n.comb. (sheaths 25–60 µm); *T. magna* (Tynni & Donner, 1980) n.comb. (sheaths 60–100 µm). Only *T. Wernadskii* is recorded in the Svanbergfjellet Formation.

Tortunema Wernadskii (Schepeleva, 1960) Butterfield, n.comb.

Figs. 24H, 27A–C

Synonymy. – □1960 *Oscillatorites Wernadskii* Schepeleva sp. nov. – Schepeleva, p. 170. □1976 *Tortunema sibirica* gen. et sp.n. – Timofeev *et al.*, p. 40, Pl. 12:2–3. □1979 *Botuobia vermiculata* Pjatiletov, sp.n. – Pjatiletov, p. 715, Pl. 1:1–4. □1980 *Oscillatoriopsis bothnica* n.sp. – Tynni & Donner, p. 15, Pl. 7:83. □1980 *Oscillatoriopsis constricta* n.sp. – Tynni & Donner, p. 15, Pl. 7:82, 85, 86. □1984 *Botuobia vermiculata* Pjatiletov, 1979 – Kolosov, pp. 43–44, Pl. 7:2. □1984 *Botuobia wernadskii* (Schep.), 1960 – Kolosov, pp. 44–46, Pl. 7:3. □1984 *Botuobia immutata* Kolosov, sp. nov. – Kolosov, p. 46, Pl. 8:1. □1988 *Tortunema cellulaefera* Pjatiletov, sp. nov. – Pjatiletov, pp. 79–80. Pl. 7:3–4. □1989 *Tortunema sibirica* Hermann, 1976, emend. Hermann – Jankauskas *et al.*, p. 123, Pl. 29:2, 4, 6, 10.

Holotype. – Slide No. 8/5122, reposited in the VNIGNI Palynological Laboratory 'sporovopyl'tsevaya laboratoriya'. Leningrad (St. Petersburg) region; borehole 3 Smerdovitsy, 211.5–381 m. Early Cambrian. Schepeleva 1960, Fig. 1.

Material. – Eight (8) specimens from shale sample 86-G-62.

Diagnosis. – A species of *Tortunema* with pseudoseptate sheaths 10–25 µm in diameter.

Description. – Filamentous sheaths, 16–27 µm wide ($\bar{x} = 20.1$ µm; s.d. = 4.3 µm; $n = 8$), with thin annular lines or thickenings. Intervals between annulations more or less regular, ca. one-half to one-third the diameter of the sheath. Disarrayed condensed cell remnants occasionally preserved within the sheath.

Discussion. – *Tortunema* was originally erected to describe septate (pseudoseptate), S-curved filaments that taper towards both ends. As with cellular trichomes, the limited terminal narrowing of these filaments is likely to be of intraspecific and/or taphonomic origin, while the curved habit is not at all constant and thus not taxonomically useful. The distinguishing features of *T. Wernadskii* n.comb. are therefore its pseudoseptate nature and size (10–25 µm in diameter; cf. Jankauskas *et al.*, 1989, p. 101 – *Botuobia wernadskii*).

Family Nostocaceae(?) Kützing, 1843

Genus *Veteronostocale* Schopf & Blacic, 1971, emend.

Synonymy. – □1971 *Veteronostocale*, n. gen. – Schopf & Blacic, p. 950. □1971 *Filiconstrictosus*, n. gen. – Schopf & Blacic, p. 947.

Type species. – *Veteronostocale amoenum* Schopf & Blacic, 1971, pp. 950–951.

Emended diagnosis. – Unbranched, uniseriate, cellular trichomes constructed of spheroidal to subspheroidal cells and having no extracellular sheath.

Discussion. – Given its *Anabaena*-like 'string of beads' appearance, *Veteronostocale* is perhaps the most likely candidate for nostocacean cyanobacteria in the Proterozoic fossil record, although comparable forms do occur in the Oscillatoriaceae (e.g., *Pseudanabaena*). The Bitter Springs form *Filiconstrictosus* is likewise much constricted at intercellular septa and is legitimately subsumed into *Veteronostocale*. The genus is represented by at least two species, *V. amoenum* and a somewhat larger form, *V. majusculum* (Schopf & Blacic, 1971) n.comb. (the latter encompassing *V. copiosus* Ogurtsova & Sergeev, 1987).

Veteronostocale amoenum Schopf & Blacic, 1971

Fig. 24I

Material. – One (1) specimen from shale sample 86-G-62.

Description. – Spheroidal to ellipsoidal cells up to 4 μm wide and 6 μm long; linked into a sinuous trichome ca. 140 μm long.

Family Scytonemataceae(?) Rabenhorst, 1863

Genus *Pseudodendron* Butterfield, n.gen.

Type species. – *Pseudodendron anteridium* n.sp.

Diagnosis. – Branched, longitudinally striate and sometimes anastomosed filamentous microfossils within an enveloping sheath; branch junctions often reinforced with a gusset of sheath material. 'Principal' axis and branches can be of markedly different diameter; individual filaments often tapered. No preserved cellularity.

Discussion. – Branching in filamentous microfossils, particularly if it is both 'false' (i.e. does not involve branching of the trichome) and anastomosing, conceivably derives from a diagenetic fusing of simple filaments. That this is not the case in *Pseudodendron* n.gen. is shown by the clearly differentiated gussets present at most branching points and its regularly recurring habit. Given its often pronounced longitudinal striation and wide range of filament diameters (within even a single specimen), *Pseudodendron* is also convincingly interpreted as having been multiseriate. Modern *Schizothrix*-type cyanobacteria are both false-branching and multiseriate and provide a good modern analogue (homologue?) for *Pseudodendron*. Earlier comparison to the rhizoids of chae-

tophoralean green algae (Butterfield *et al.* 1988) now appears untenable.

Striate, probably multiseriate filaments have previously been reported from Proterozoic fossil assemblages, but apart from one unnamed form (Nyberg & Schopf 1984; see below) these differ substantially from *Pseudodendron* n.gen. *Eomicrocoleus crassus* Horodyski & Donaldson, 1980, and *Manicosiphoninema shuiyouense* Yan, 1992, have an outer sheath but appear not to branch, while *Talakania obscura* Kolosov, 1984, branches, but has no sheath, and is cellular. Clearly false-branched and tapered filaments are characteristic of Paleoproterozoic *Changchengonema densa* Yan, 1989, but these are not obviously striated and do not exhibit sheath-reinforced branch junctions; likewise, *Eoholynia mosquensis* Gnilovskaya, 1975, which, in addition, bears multiple 'sporangia' and is considerably larger than *Pseudodendron*. Finally, although the written diagnosis of *Ulophyton rifeicum* Timofeev & Hermann, 1979, coincides broadly with that of *Pseudodendron*, the type material of this putative metaphyte is clearly unrelated and is possibly not even of primary origin. A second species, *U. longiscapus* Hermann, 1989 (*in* Jankauskas *et al.* 1989), is unquestionably biogenic and may be a species of *Pseudodendron*.

Pseudodendron n.gen. exhibits a wide but relatively continuous range of form in the Svanbergfjellet shales and is here considered to comprise a single species, *P. anteridium* n.sp. Branched filaments with an overall similar habit but no evidence of a differentiated envelope or longitudinal striation are classified as *Pseudodendron* sp. (e.g., Fig. 23I).

Etymology. – From the Greek *pseudes* – false, and *dendron* – tree, with reference to the false-branching habit.

Pseudodendron anteridium Butterfield, n.sp.

Fig. 28A–G, J

Synonymy. – □1984 Branched filamentous structure – Nyberg & Schopf, p. 770, Fig. 11F.

Holotype. – HUPC 62720, Fig. 28J; Slide P-2945-7M, England-Finder coordinates L-41-3.

Type locality. – Lower Dolomite Member, Svanbergfjellet Formation, Polarisbreen (79°10'N, 18°12'E); 17 m below the base of the *Minjaria* biostrome.

Material. – One hundred seven (107) specimens: 9 in bedding-parallel thin section; 98 isolated by acid maceration (some as extensive mats). From shale samples P-2945 and 86-G-62. Eight (8) designated paratypes: HUPC 62721, Fig. 21A; HUPC 62899–62906, Fig. 28A–I.

Diagnosis. – A species of *Pseudodendron* with maximum filament diameter less than 60 μm.

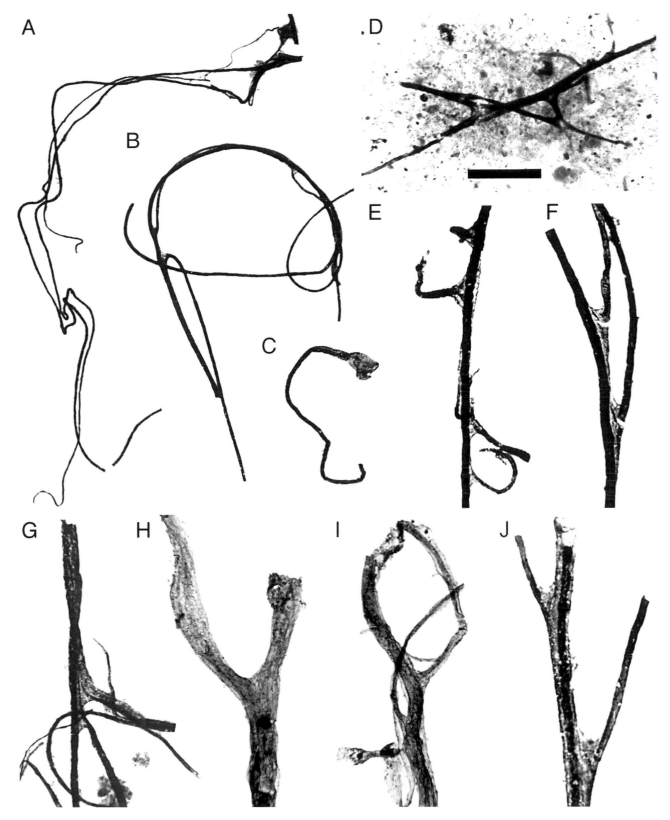

Fig. 28. Pseudodendron anteridium n.gen., n.sp. From shales of the Lower Dolomite Member, Polarisbreen (A–G, J), and the Algal Dolomite Member, Geerabukta (H, I). Scale bar in D equals 220 μm for A–C; 50 μm for D, H–J; 80 μm for E–G. □A. HUPC 62899; P-2945-9M (N-29-0). □B. HUPC 62900; P-2945-4M (N-34-3). □C. HUPC 62901; P-2945-13M (O-30-0); with terminal swelling. □D. HUPC 62902; 86-G-62-33 (R-49-4); illustrating the capacity to fuse separated axes. □E. HUPC 62903; P-2945-5M (K-29-2). □F. HUPC 62904; P-2945-74M (L-18-4). □G. P-2945 (destroyed). □H. HUPC 62905; 86-G-62-96M (L-39-2); envelope just discernable at the branch junction. □I. HUPC 62906; 86-G-62-237M (K-17-0); detail of a 1.25 mm long filament; note the thin enveloping sheath and three branches. □J. HUPC 62720; P-2945-7M (L-41-3); holotype.

Description. – Heterogeneously branched, often tapered filaments, 3–60 μm wide and up to 2.7 mm long, usually with a 1–10-μm-scale longitudinal striation. Outer sheath present but not always obvious; conspicuous at branch junctions where it can occur on the inside angle as a prominent subtriangular gusset. At least two levels of branching present; separate axes sometimes grown back together. Rare filaments with a terminal expansion. Overall habit as either isolated specimens or matted associations up to 2.2 mm in maximum dimension. Matted forms tend to be split (mechanically?) along the longitudinal striae.

Discussion. – *Pseudodendron anteridium* n.sp. is a conspicuous component of the Lower Dolomite Member shale assemblage (P-2945), with specimens recovered as single, relatively nondescript filaments; sparsely to regularly branched forms; and filaments terminating (or originating?) in a bulbous expansion (Fig. 28C). In the Upper Algal Dolomite shale it is less common but occasionally occurs in relatively extensive mats; in one instance it clearly exhibits its capacity to rejoin separated axes (Fig. 28D).

That the branching of *P. anteridium* n.sp. is 'false' is indicated by the confluence of lateral axes into a primary axis without their actual incorporation into its central 'trichome'; rather, they maintain their individuality and continue on in parallel formation within the outer sheath of the main filament (Fig. 28G). Like the outer envelope, the filament cores appear to be composed of extracellular sheath material, as indicated in their occasional confluence with the primary envelope (Fig. 28E); actual trichomes appear not to be preserved.

Svanbergfjellet *P. anteridium* n.sp. compare closely with silicified fossils described from the 680–790 Ma old Min'yar Formation of the southern Urals (Nyberg & Schopf 1984). These similarly multiseriate, ensheathed, and false-branched filaments contain up to 36 narrow (2.0–5.0 μm), tubular sheaths.

Etymology. – From the Greek, *anteridion* – support, buttress, with reference to the often conspicuous reinforcement of the branch junctions with sheath material.

Incertae Sedis

Genus *Brachypleganon* Lo, 1980

Type species. – *Brachypleganon khandanum* Lo, 1980, p. 156.

Discussion. – Small, rod-shaped Proterozoic microfossils have been assigned to *Eosynechococcus* Hofmann, 1976, *Archaeoellipsoides* Horodyski & Donaldson, 1980, *Brachypleganon* Lo, 1980, and *Bactrophycus* Zhang, 1985. On the basis of an apparently continuous morphological gradient Golovenoc & Belova (1984) subsumed *Archaeoellipsoides* into *Eosynechococcus* and erected a number of new species

such that the taxon now includes forms with an aspect ratio as high as 3.6:1 (see *Eosynechococcus*, above). By contrast, *Brachypleganon* has a length-to-width ratio of ca. 7:1. (The 4.0–15:1 aspect of *Bactrophycus* suggests that it may be a junior synonym of *Brachypleganon*.)

Brachypleganon thus stands as a legitimate form genus. More interestingly, an appreciation of its preserved behavior also supports its distinction as a natural taxon. Unlike the end-to-end linked strings or closely packed aggregations typical of *Eosynechococcus* (Hofmann 1976), *Brachypleganon* usually occurs as loosely aggregated and oriented, unattached colonies (Lo 1980); within the form, this habit is at least as telling as measured differences in aspect ratio. Svanbergfjellet *Brachypleganon* are preserved in shales rather than the silicified carbonate of the type material; however, both the individual rod morphology and the colonial organization are closely comparable.

Brachypleganon khandanum Lo, 1980
Fig. 22J–K

Material. – Eighteen (18) populations from shale sample 86-G-62.

Description. – Rod-shaped microfossils, 1.5 μm wide by 6–16 μm long ($\bar{x} = 10.1$ μm; s.d. = 2.2 μm; $n = 150$); ca. 7 times longer than wide. Ends rounded but not significantly tapered. Envelope absent. Usually occurring in loosely oriented populations (colonies) of up to ca. 100 individuals.

Discussion. – The combination of rod shape and oriented colonies was earlier taken as implying a heterotrophic physiology for these fossils (Butterfield *et al.* 1988). These are now considered as insufficient to rule out a possible photosynthetic metabolism, and *B. khandanum* is consequently classified as *incertae sedis*.

Genus *Chlorogloeaopsis* Maithy, 1975

Type species. – *Chlorogloeaopsis zairensis* Maithy, 1975, p. 139.

Discussion. – In addition to *Chlorogloeaopsis*, spheroidal microfossils arrayed into filament-like colonies are found in *Polysphaeroides* Hermann, 1976, *Cyanothrixoides* Golovenoc & Belova, 1985, and *Zinkovioides* Hermann, 1985. *Chlorogloeaopsis* differs from the type species of *Polysphaeroides* (*P. filliformis*) in having no enveloping sheath.

The original spelling of this genus was '*Cholorogloeaopsis*', although it was erected with reference to the modern cyanobacterium *Chlorogloea* and was occasionally spelled '*Chlorogloeaopsis*' in the accompanying discussion (Maithy 1975). The spelling is here corrected in accordance with ICBN Article 73.1.

Chlorogloeaopsis zairensis Maithy, 1975

Fig. 20I

Synonymy. – □1975 *Cholorogloeaopsis* [*sic*] *zairensis* sp. nov. – Maithy, p. 139, Pl. 3:21–23. □1985 *Polysphaeroides biseritus* Liu (sp. nov.) – Xing *et al.*, p. 65, Pl. 7:14–15.

Material. – One (1) specimen from shale sample 86-G-62.

Description. – Spheroidal to moderately ellipsoidal cells forming an elongate colony ca. 15 μm (2 cells) wide and 365 μm long. Maximum cell dimensions 7–13 μm ($\bar{x}$ = 9.6 μm; s.d. = 1.6 μm, n = 30). Cells typically with a single dark inclusion. Terminal cells unordered. Enveloping sheath absent.

Discussion. – Svanbergfjellet *C. zairensis* correspond closely to the type material from the Upper Group of the Bushimay System, Zaire, as well as to synonymous *P. biseritus* from China.

Genus *Digitus* Pjatiletov, 1980

Type species. – *Digitus fulvus* Pjatiletov, 1980, p. 68 (*in* Pjatiletov & Karlova 1980).

Discussion. – Both *Digitus* Pjatiletov, 1980, and *Brevitrichoides* Jankauskas, 1980(b), are form taxa encompassing smooth-walled filamentous microfossils with both ends intact (i.e. entire filaments). Both might be described as 'short filaments' or 'long rods', and the absolute and relative (aspect ratio) dimensions of the two overlap. If a generic-level distinction were to be made it would be in the overall narrower aspect and the slightly more tapered ends of *Digitus*; this is also the habit of the Svanbergfjellet forms.

Digitus adumbratus Butterfield, n.sp.

Fig. 7H–I

Holotype. – HUPC 62719, Fig. 7I; Slide 86-G-62-133M, England-Finder coordinates N-29-2.

Type locality. – Algal Dolomite Member, Svanbergfjellet Formation, Geerabukta (79°35'30"N, 17°44'E); 55 m above base of member.

Material. – Five (5) specimens from shale sample 86-G-62. One (1) designated paratype: HUPC 62731, Fig. 7H.

Diagnosis. – A species of *Digitus* ca. 35 μm wide and 250–500 μm long. Surface shagrinate.

Description. – Rod-shaped entire filaments, terminally tapered for ca. 10% of their length; rounded ends. Surface shagrinate but otherwise unornamented. Width, 34–40 μm ($\bar{x}$ = 36 μm, S.D = 2 μm, n = 5); length, 293–468 μm ($\bar{x}$ = 373 μm, s.d. = 30 μm, n = 5).

Discussion. – *Digitus adumbratus* n.sp. falls entirely outside the size range of previously named species of *Digitus* (maximum width and length of *D. fulvus* are, respectively, 25 and 125 μm) and therefore warrants separate species status. The shagrinate surface texture of *D. adumbratus* is also distinctive and likely represents the remnants of a mucilaginous envelope.

Etymology. – From the Latin *adumbratus*, vaguely outlined, overshadowed, with reference to the shagrinate surface texture.

Genus *Myxococcoides* Schopf, 1968

Type species. – *Myxococcoides minor* Schopf, 1968, p. 676.

Myxococcoides minor Schopf, 1968

Fig. 20C

Material. – Five (5) colonies and numerous isolated individuals in chert sample P-2664.

Description. – Spheroidal microfossils, 8–18 μm in diameter, commonly aggregated into loose colonies of several to ca. 20 cells. Colonies typically embedded in a thin organic matrix.

Discussion. – These populations compare closely with the type *Myxococcoides* populations, except that their size range spans that of all three of the originally named Bitter Springs species (even within a single colony). In lieu of a major revision of small spheroidal microfossils (which would include at least 28 described species of *Myxococcoides*) this population is assigned to *M. minor*, with the suggestion that the size range of the taxon be altered to include cells 8–18 μm in diameter.

Myxococcoides cantabrigiensis Knoll, 1982

Fig. 20F, J

Material. – Abundant colonies in chert sample 86-G-14; other scattered occurrences.

Description. – Relatively thick-walled spheroidal microfossils, 11–20 μm in diameter, typically clustered into loose colonies of several to several tens of cells. Envelope and matrix absent.

Myxococcoides spp.

Fig. 20G

Discussion. – A variety of smooth-walled spheroidal microfossils occur in silicified carbonates of the Svanbergfjellet

Formation. Most are thin-walled, colonial (but without an enclosing matrix), and occur within dense, dominantly filamentous microbial mats. They range from 8 to 50 μm in diameter (typically 10–30 μm) and are here assigned to *Myxococcoides* spp. Also included are spheroids with a single envelope (Fig. 20G) but otherwise lacking features diagnostic of *Gloeodiniopsis* (see above).

Genus *Ostiana* Hermann, 1976

Type species. – *Ostiana microcystis* Hermann, 1976, p. 43 (*in* Timofeev *et al.* 1976).

Ostiana microcystis Hermann, 1976
Fig. 5F–I

Material. – Sixty nine (69) cohesive colonies: 42 in bedding-parallel thin section; 27 isolated by acid maceration. An additional 63 loose colonies or associations of similar cells also recorded in thin section . From shale samples 86-G-62 and 86-G-61.

Description. – Closely packed sheets to loose associations of spheroidal cells, one or (rarely) two layers thick; up to 2100 μm broad. Cells commonly deformed due to mutual compression and often with dark inclusions. Cell diameter 11–27 μm with a prominent mode at 16 μm. Extracellular matrix absent.

Discussion. – Svanbergfjellet *O. microcystis* isolated by acid maceration occur in three distinct habits that may at some time warrant separate taxonomic designations: (1) tightly packed, polyhedral and inclusion-bearing cells that compare well with the type material (Fig. 5G–H); (2) closely packed but undistorted thin-walled cells arranged in bi-layered sheets (Fig. 5I); and (3) more loosely packed sheets of undistorted thicker-walled cells (Fig. 5F). The first form is also abundant in bedding-parallel thin section where a continuum from fully cohesive sheets to localized associations of isolated cells can be observed. Thus, while *O. microcystis* appears superficially to be multicellular – the sheets do act as a unit, wrinkling and folding upon themselves (Fig. 5G, H), and in several instances include seemingly structured apertures (Fig. 5G) – it clearly represents a simple pluricellular (*sensu* Awramik & Valentine 1985) construct of aggregated or unseparated unicells.

Monostromatic cellular colonies are characteristic of a number of extant cyanobacterial taxa, including *Microcrocis*, *Holopedia* and *Merismopedium* (Frank & Landman 1988). As with *O. microcystis*, the cells of these taxa may be close-packed and polygonally deformed, and/or less closely associated such that peripheral cells are often separated from the main colony. They differ markedly from Svanbergfjellet *Ostiana* in their much smaller cell size (2.5–6.8 μm) and by the presence of a thin mucous layer around the colony (such a layer does occasionally occur in the type material of *Ostiana*). *Ostiana microcystis* was almost certainly photosynthetic, but with its relatively large cells it is not immediately obvious whether it was prokaryotic or eukaryotic.

Genus *Palaeosiphonella* Licari, 1978

Type species. – *Palaeosiphonella cloudii* Licari, 1978, p. 788.

Palaeosiphonella sp.
Fig. 21E, H–J

Material. – Abundant in silicified carbonate sample P-3075.

Description. – Gregarious, more or less vertically oriented, and commonly branched tubes, 14–40 μm in diameter ($\bar{x} = 22$ μm; s.d. = 6 μm; $n = 25$); moderately to highly sinuous, often with localized constrictions and/or expansions. Walls dark and particulate, 1–2 μm thick. At least two orders of branching present.

Discussion. – Unlike the type species of *Palaeosiphonella*, these branched tubular fossils are vertically oriented, imposing a conspicuous fabric to a single, ca. 2 mm thick laminae of a silicified, flat-laminated carbonate of the Lower Limestone Member. Moreover, the walls of the Svanbergfjellet tubes are markedly sinuous and have a dark, particulate construction. This latter feature is clearly taphonomic, but it is interesting to note that immediately associated spheroidal microfossils exhibit the translucent, psilate walls typical of well-preserved chert biotas elsewhere, and of *P. cloudii*. Nonetheless, the overall form and size of these branched filamentous fossils are broadly comparable to those of *Palaeosiphonella* and warrant their inclusion in that form-genus; species designation awaits more and better preserved material.

In the absence of preserved cellular structure, the source of these fossils is speculative. The default, though not necessarily correct, interpretation is that they represent false branching prokaryotes comparable to modern scytonematacean cyanobacteria, as is likely the case for *Ramivaginalis* Nyberg & Schopf, 1984, and *Palaeosiphonella cloudii*. Equally plausible comparisons, however, can be drawn with certain eukaryotic algae, or even metazoan traces. Vertically oriented, but unwalled and less sinuous and branched filamentous structures are preserved as 13–60 μm wide casts in the immediately overlying Draken Formation (Knoll *et al.* 1991).

Filament-bearing body
Fig. 21B

Material. – One (1) specimen from shale sample 86-G-62.

Description. – Dark central carbonaceous body, 105×35 µm, bearing several radially oriented, fibrous filaments; filaments up to 400 µm long and distally expanded (13–30 µm wide).

Discussion. – This unique specimen appears to differentiate two distinct cell or tissue types, thereby suggesting a relatively complex grade of organization. Conversely, it might be interpreted as two quite unrelated structures that were fused through sedimentary compaction. Both its regular radiating pattern and the uniqueness of the two components argue against this latter alternative.

Sub-vertical branched(?) filaments(?) in apatite

Fig. 23F–G

Material. – Common in nodular apatite samples SV-2 and SV-3.

Description. – Gregarious, often branched(?) sub-vertical filamentous structures, 8–35 µm in diameter.

Discussion. – Abundant, poorly preserved filamentous structures are concentrated along discrete laminae in shale-hosted apatite nodules of the Lower Dolomite Member. Their vertical orientation and apparent branching suggest that they are biogenic, but further taxonomic assessment is unfeasible; the 'branching' may be a product of diagenesis. As these nodules were formed before significant compaction of the surrounding shale, they provide a unique view of the vertical dimension in a Neoproterozoic muddy environment.

Mineralized filaments – goethite(?)

Fig. 26C–F

Material. – Abundant examples in shale sample 86-G-30.

Description. – Solitary to thickly amassed sinuous 'filaments' preserved three-dimensionally as mineral overgrowths (goethite?) on *Siphonophycus*-like sheaths, or entirely mineralic. Diagenetic fusion of intersecting filaments common. Diameter variable but less than 25 µm.

Discussion. – Mineralized filaments are found on exposed bedding planes (Fig. 26C) and in bedding-parallel thin sections (Fig. 26D–F) of a single Algal Dolomite Member shale (sample 86-G-30). That they originated as biological structures is evident from both their pronounced matting habit (Fig. 26F) and the occasional preservation of an internal organic-walled sheath (Fig. 26D). It is also clear, however, that much of the morphology of these fossils is secondary; it is the mineral overgrowth that accounts for most of their girth, and that has fused separate filaments to give them the appearance of branching (Fig. 26E–F).

The yellow mineral defining these fossils is translucent, anisotropic, and iron-rich (EDAX analysis), possibly goethite. In addition to its early precipitation on and within filamentous organisms, it occurs as small veins and localized crystals throughout the sample, a habit identifying the filament rinds as a product of early diagenesis rather than primary biomineralization.

Possible eukaryotic algal filaments reported from the Late Proterozoic Chuar Group by Horodyski & Bloeser (1983) exhibit a number of features reminiscent of the Svanbergfjellet fossils. Although somewhat larger (40–130 µm wide), these filaments likewise occur three-dimensionally on shale bedding planes, exhibit a sinuous and pseudobranching habit, and sometimes preserve an internal cylindrical structure (original sheath?). It is therefore possible that they, too, represent mineralic overgrowths of considerably smaller *Siphonophycus*-like filaments and as such are less convincingly interpreted as eukaryotic.

Acknowledgements – We would like to thank Roger Buick, Stephen Grant, and Julian Green for numerous suggestions and ongoing encouragement. Gonzalo Vidal thoroughly reviewed, corrected, and commented on the manuscript, Harvey Marchant (Australian Antarctic Division) provided the SEM of modern *Coelastrum*, and Carl Mendelson (Beloit College) helped in obtaining obscure literature; we gratefully acknowledge the time and energy of Stefan Bengtson in his role as editor of *Fossils and Strata*. Field and laboratory work was supported by NSF grants DPP 85-15863 (to AHK and KS) and BSR 90-17747 (to AHK), and publication costs were generously provided for by the Norwegian Science Research Council. NJB was supported by a Natural Sciences and Engineering Research Council of Canada post-graduate fellowship at Harvard University and a Trevelyan Research Fellowship, Selwyn College, Cambridge.

References

Alexander, M. 1973: Nonbiodegradable and other recalcitrant molecules. *In: Biotechnology and Bioengineering, Vol. 15.* 611–647. John Wiley & Sons, Inc.

Allison, C.W. & Awramik, S.M. 1989: Organic-walled microfossils from earliest Cambrian or latest Proterozoic Tindir Group rocks, northwest Canada. *Precambrian Research 43*, 253–294.

Allison, P.A. 1988: *Konservat-Lagerstätten*: cause and classification. *Paleobiology 14*, 331–344.

Amard, B. 1992: Ultrastructure of *Chuaria* (Walcott) Vidal and Ford (Acritarcha) from the late Proterozoic Pendjari Formation, Benin and Burkina-Faso, west Africa. *Precambrian Research 57*, 121–133.

Asmerom, Y., Jacobsen, S.B., Knoll, A.H., Butterfield, N.J. & Swett, K. 1991: Strontium isotopic variations of Neoproterozoic seawater: Implications for crustal evolution. *Geochimica et Cosmochimica Acta 55*, 2883–2894.

Assejeva, E.A. 1982: Novye vidy planktonnykh vodoroslei venda Volyno-Podolii. [New species of planktonic algae from the Vendian of Volyn-Podoli.] *In* Teslenko, Y.V. (ed.): *Sistematika i Evoliutsiya Drevnikh Rastenij Ukrainy*, 5–16. Naukova Dumka, Kiev.

Assejeva, E.A. 1983: Stratigraficheskoe znachenie pozdnedokembrijskikh mikrofossilij iugo-zapada Vostochno-Evropejskoj Platformy. [Stratigraphic significance of late Precambrian microfossils from the southwestern East European Platform.] *In* Riabenko, V.A. (ed.): *Stratigrafiya i Formatsii Dokembriya Ukrainy*, 148–176. Naukova Dumka, Kiev.

Assejeva, E.A. & Velikanov, V.A. 1983: Novaya nakhodka iskopaemykh fitoostatkov v liadovskikh sloyakh venda Podolii (verkhnij dokembrij). [New finds of fossil plant remains in the Liada beds of the Vendian of Podoli (upper Precambrian)]. *In* Vialov, O.S. (ed.): *Iskopaemaya Fauna i Flora Ukrainy*, 3–8. Naukova Dumka, Kiev.

Atkinson, A.W.J., Gunning, B.E.S. & John, P.C.L. 1972: Sporopollenin in the cell wall of *Chlorella* and other algae: ultrastructure, chemistry, and incorporation of ¹⁴C-acetate, studied in synchronous cultures. *Planta 107*, 1–32.

Avnimelech, Y., Troeger, B.W. & Reed, L.W. 1982: Mutual flocculation of algae and clay: evidence and implications. *Science 216*, 63–65.

Awramik, S.M. & Valentine, J.W. 1985: Adaptive aspects of the origin of autotrophic eukaryotes. *In* Tiffney, B.H. (ed.): *Geological Factors and the Evolution of Plants*. 11–21. Yale University Press, New Haven.

Bauld, J. 1981: Geobiological role of cyanobacterial mats in sedimentary environments: production and preservation of organic matter. *BMR Journal of Australian Geology and Geophysics 6*, 307–317.

Berkaloff, C., Casadevall, E., Largeau, C., Metzger, P., Peracca, S. & Virlet, J. 1983: The resistant polymer of the walls of the hydrocarbon-rich alga *Botryococcus braunii. Phytochemistry 22*, 389–397.

Bhattacharya, D., Elwood, H.J., Goff, L.J. & Sogin, M.L. 1990: Phylogeny of *Gracilaria lemaneiformis* (Rhodophyta) based on sequence analysis of its small subunit ribosomal RNA coding region. *Journal of Phycology 26*, 181–186.

Birch, P.B., Gabrielson, J.O. & Hamel, K.S. 1983: Decomposition of *Cladophora* I. Field studies in the Peel-Harvey estuarine system, Western Australia. *Botanica Marina 26*, 165–171.

Bloeser, B. 1985: *Melanocyrillium*, a new genus of structurally complex late Proterozoic microfossils from the Kwagunt Formation (Chuar Group), Grand Canyon, Arizona. *Journal of Paleontology 59*, 741–765.

Bonner, J.T. 1988: *The Evolution of Complexity by Means of Natural Selection*. 260 pp. Princeton University Press, Princeton.

Børgesen, F. 1913: The marine algae of the Danish West Indies. Part I, Chlorophyceae. *Dansk Botanisk Arkiv 1*. 158 pp.

Bowman, B.H., Taylor, J.W., Brownlee, A.G., Lee, J., Lu, S.-D. & White, T.J. 1992: Molecular evolution of the fungi: relationship of the basidiomycetes, ascomycetes, and chytridiomycetes. *Molecular Biology and Evolution 9*, 285–296.

Briggs, D.E.G. & Williams, S.H. 1981: The restoration of flattened fossils. *Lethaia 14*, 157–164.

Brotzen, F. 1941: Några bidrag till visingsöformationens stratigrafi och tektonik. *Geologiska Föreningens Förhandlingar 63*, 245–261.

Bruns, T.D., White, T.J. & Taylor, J.W. 1991: Fungal molecular systematics. *Annual Review of Ecology and Systematics 22*, 525–564.

Bu Déan 1985: Discovery of Cyanophycean filamentous microfossils from the Ganjingzi Formation (Late Precambrian) of the southern Liaodong Peninsula. *In: Selected Papers from the 1th National Fossil Algal Symposium*, 207–212. Geological Publishing House, Beijing.

Burns, R.G. 1979: Interaction of microorganisms, their substrates and their products with soil surfaces. *In* Ellwood, D.C., Melling, J. & Rutter, P. (eds.): *Adhesion of Microorganisms to Surfaces*, 109–138. Academic Press, London.

Buss, L.W. 1987: *The Evolution of Individuality*. 201 pp. Princeton University Press, Princeton.

Butterfield, N.J. 1990: Organic preservation of non-mineralizing organisms and the taphonomy of the Burgess Shale. *Paleobiology 16*, 272–286.

Butterfield, N.J. & Chandler, F.W. 1992: Palaeoenvironmental distribution of Proterozoic microfossils, with an example from the Agu Bay Formation, Baffin Island. *Palaeontology 35*, 943–957.

Butterfield, N.J. & Rainbird, R.H. 1988: The paleobiology of two Proterozoic shales. *Geological Society of America, Abstracts with Programs 20*, A103.

Butterfield, N.J., Knoll, A.H. & Swett, K. 1988: Exceptional preservation of fossils in an Upper Proterozoic shale. *Nature 334*, 424–427.

Butterfield, N.J., Knoll, A.H. & Swett, K. 1990: A bangiophyte alga from the Proterozoic of arctic Canada. *Science 250*, 104–107.

Cavalier-Smith, T. 1978: Nuclear volume control by nucleoskeletal DNA, selection for cell volume and cell growth rate, and the solution of the DNA C-value paradox. *Journal of Cell Science 34*, 247–278.

Chalansonnet, S., Largeau, C., Casadevall, E., Berkaloff, C., Peniguel, G. & Couderc, R. 1987: Cyanobacterial resistant biopolymers. Geochemical implications of the properties of *Schizothrix* sp. resistant material. *Organic Geochemistry 13*, 1003–1010.

Chapman, F. 1935: Primitive fossils, possibly atrematous and neotrematous Brachiopoda, from the Vindhyans of India. *Records of the Geological Survey of India 69*, 109–120.

Chen Menge & Liu Kuiwu 1986: The Geological significance of newly discovered microfossils from the upper Sinian (Doushantuo age) phosphorites. *Scientia Geologica Sinica 1986*, 46–53.

Chen Menge & Xiao Zongzheng 1991: Discovery of the macrofossils in the upper Sinian Doushantuo Formation at Miaohe, eastern Yangtze Gorges. *Scientia Geologica Sinica 1991*, 317–324.

Conway Morris, S. 1979: Middle Cambrian polychaetes from the Burgess Shale of British Columbia. *Philosophical Transactions of the Royal Society of London B 285*, 227–274.

Conway Morris, S. 1986: The community structure of the Middle Cambrian Phyllopod Bed (Burgess Shale). *Palaeontology 29*, 423–467.

Cookson, I. C. & Eisenack, A. 1960: Microplankton from Australian Cretaceous sediments. *Micropaleontology 6*, 1–18.

Copeland, J.J. 1936: Yellowstone thermal Myxophyceae. Annals of the New York Academy of Science 36, 1–232.

Dawson, M.P., Humphrey, B.A. & Marshall, K.C. 1981: Adhesion: a tactic in the survival strategy of a marine vibrio during starvation. *Current Microbiology 6*, 195–199.

Deflandre, G. 1937: Microfossiles des silex crétacés. Deuxième Partie. *Annales de Paléontologie 26*, 51–103.

Derry, L.A., Keto, L.S., Jacobsen, S.B., Knoll, A.H. & Swett, K. 1989: Sr isotopic variations in Upper Proterozoic carbonates from Svalbard and East Greenland. *Geochimica et Cosmochimica Acta 53*, 2331–2339.

Downie, C. 1982: Lower Cambrian acritarchs from Scotland, Norway, Greenland, and Canada. *Transactions of the Royal Society of Edinburgh: Earth Sciences 72*, 257–285.

Drugg, W. S. 1967: Palynology of the upper Moreno Formation (Late Cretaceous – Paleocene), Escarpado Canyon, California. *Palaeontographica B 120*, 1–71.

Du Huiying. 1985: Late Precambrian microflora from northern slope of the Qinling Range and its stratigraphic significance. *In: Selected Papers from the 1th National Fossil Algal Symposium*, 155–168. Geological Publishing House, Beijing.

Du Rulin 1982: The discovery of the fossils such as *Chuaria* in the Qingbaikou system in northwestern Hebei and their significance. *Geology Review (Beijing) 28*, 1–7.

Du Rulin, Tian Lifu & Li Hanbang 1986: Discovery of megafossils in the Gaoyuzhuang Formation of the Changchengian System, Jixian. *Acta Geologica Sinica 60*, 115–120.

Duan Cheng-hua 1982: Late Precambrian algal megafossils *Chuaria* and *Tawuia* in some areas of eastern China. *Alcheringa 6*, 57–68.

Eisenack, A. 1931: Neue Mikrofossilien des baltischen silurs. I. *Paläontologische Zeitschrift 13*, 74–118.

Eisenack, A. 1951: Über Hystrichosphaerideen und andere Kleinformen aus Baltischem Silur und Kambrium. *Senckenbergiana Lethaea 32*, 187–204.

Eisenack, A. 1955: Chitinozoen, Hystrichosphären un andere Mikrofossilien aus dem Beyrichia-Kalk. *Senckenbergiana Lethaea 36*, 157–188.

Eisenack, A. 1958: *Tasmanites* Newton 1875 n.g. als Gattungen der Hystrichosphaeridea. *Palaeontographica A 110*, 1–19.

Eisenack, A. 1966: Über *Chuaria wimani* Brotzen. *Neues Jahrbuch für Geologie und Paläontologie Monatshefte 1966*, 52–56.

Eisenack, A. 1969: Zur Systematik einiger paläozoischer Hystrichosphären (Acritarcha) des baltischen Gebietes. *Neues Jahrbuch für Geologie und Paläontologie Abhandlungen 133*, 245–266.

Evitt, W.R. 1963a: Occurrence of freshwater alga *Pediastrum* in Cretaceous marine sediments. *American Journal of Science 261*, 890–893.

Evitt, W.R. 1963b: A discussion and proposals concerning fossil dinoflagellates, hystrichospheres and acritarchs, II. *Proceedings of the National Academy of Sciences USA 49*, 298–302.

[Fairchild, T. 1975: The geological setting and paleobiology of a late Precambrian stromatolitic microflora from South Australia. Unpublished Ph.D. Dissertation. University of California, Los Angeles. 272 pp.]

Fallon, R.D. & Brock, T.D. 1979: Decomposition of blue-green algal (cyanobacterial) blooms in Lake Mendota, Wisconsin. *Applied and Environmental Microbiology 37*, 820–830.

Farmer, J.D. 1992: Grazing and bioturbation in modern microbial mats. *In* Schopf, J.W. & Klein, C. (eds.): *The Proterozoic Biosphere, A Multidisciplinary Study,* 295–297. Cambridge University Press, Cambridge.

Fattom, A. & Shilo, M. 1984: Hydrophobicity as an adhesion mechanism of benthic cyanobacteria. *Applied and Environmental Microbiology 47*, 135–143.

Ferris, F.G., Fyfe, W.S. & Beveridge, T.J. 1988: Metallic ion binding by *Bacillus subtilis*: Implications for the fossilization of microorganisms. *Geology 16*, 149–152.

Fontes, D.E., Mills, A.L., Hornberger, G.M. & Herman, J.S. 1991: Physical and chemical factors influencing transport of microorganisms through porous media. *Applied and Environmental Microbiology 57*, 2473–2481.

Ford, T.D. & Breed, W.J. 1973: The problematical Precambrian fossil *Chuaria. Palaeontology 16*, 535–550.

Foree, E.G. & McCarty, P.L. 1970: Anaerobic decomposition of algae. *Environmental Science and Technology 4*, 842–849.

Frank, T.H. & Landman, A.G. 1988: Morphology and taxonomy of *Microcrocis* sp. (Chroococcales, Cyanophyta) from Lake Vechten, The Netherlands, with a note on its observed movements. *Archiv für Hydrobiologie, Algological Studies 49*, 397–410.

Fritsch, F.E. 1935: *The structure and Reproduction of the Algae, Volume 1.* 791 pp. Macmillan, New York.

Fu Jun-hui 1989: New materials of late Precambrian Huainan biota fossil in Souxian, Anhui. *Acta Palaeontologica Sinica 28*, 72–78.

Geitler, L. 1925: Cyanophyceae. *In* Pascher, A. (ed.): *Die Süsswasserflora Deutschlands, Österreichs, und der Schweiz, Volume 12 (Cyanophyceae, Cyanochloridinae, Chlorobacteriaceae),* 1–450. Gustav Fischer, Jena.

Gnilovskaya, M.B. 1975: Novye dannye o prirode vendotenid. [New data on the nature of the vendotaenids.] *Doklady Akademii Nauk SSSR 221*, 953–955.

Gnilovskaya, M.B. (ed.) 1988: *Vendotenidy Vostochno-Evropejskoj Platformy.* [*Vendotaenids of the East European Platform.*] 143 pp. Nauka, Leningrad.

Golovenoc, V.K. & Belova, M.Y. 1984: Rifejskie mikrobioty v kremnyakh iz billyakhskoj serii Anabarskogo podniatiya. [Riphean microbiota in cherts of the Billyakh Group on the Anabar Uplift. *Paleontological Journal 1984:4,* 20–30.] *Paleontologicheskij Zhurnal 1984:4),* 23–32.

Golovenoc, V.K. & Belova, M.Y. 1985: Rifejskie mikrobioty v kremnyakh Yenisejskogo kryaza. [Riphean microbiotas in cherts of the Yenisej Ridge. *Paleontological Journal 1985:2,* 88–99.] *Paleontologicheskij Zhurnal 1985:2,* 94–103.

Golub, I.N. 1979: Novaya gruppa problematichnykh mikroobrazovanij v vendskikh otlozheniyakh Orshanskoj vpadiny (Russkaya platforma). [A new group of problematic microstructures in Vendian deposits of the Orshanka Basin (Russian Platform).] *In* Sokolov, B.S. (ed.): *Paleontologiya Dokembriya i Rannego Kembriya.* 147–155. Nauka, Leningrad.

Golubic, S. & Barghoorn, E.S. 1977: Interpretation of microbial fossils with special reference to the Precambrian. *In* Flügel, E. (ed.): *Fossil Algae.* 1–14. Springer, Berlin.

Golubic, S. & Hofmann, H.J. 1976: Comparison of Holocene and mid-Precambrian Entophysalidaceae (Cyanophyta) in stromatolitic algal mats: cell division and degradation. *Journal of Paleontology 50,* 1074–1082.

Grant, S.W.F., Knoll, A.H. & Germs, G.J.B. 1991: Probable calcified metaphytes in the latest Proterozoic Nama Group, Namibia: origin, diagenesis and implications. *Journal of Paleontology 65,* 1–18.

Gray, J. 1960: Fossil chlorophycean algae from the Miocene of Oregon. *Journal of Paleontology 34,* 453–463.

Green, J. W., Knoll, A. H., Golubic, S. & Swett, K. 1987: Paleobiology of distinctive benthic microfossils from the upper Proterozoic Limestone–Dolomite 'Series', central East Greenland. *American Journal of Botany 74,* 928–940.

Green, J.W., Knoll, A.H. & Swett, K. 1989: Microfossils from silicified stromatolitic carbonates of the Upper Proterozoic Limestone–Dolomite 'Series', central East Greenland. *Geological Magazine 126,* 567–585.

Grey, K. & Williams, I.R. 1990: Problematic bedding-plane markings from the Middle Proterozoic Manganese Subgroup, Bangemall Basin, Western Australia. *Precambrian Research 46,* 307–328.

Gunnison, D. & Alexander, M. 1975: Resistance and susceptibility of algae to decompostion. *Limnology and Oceanology 20,* 64–70.

Gussow, W.C. 1973: *Chuaria* sp. cf. *C. circularis* Walcott from the Precambrian Hector Formation, Banff National Park, Alberta, Canada. *Journal of Paleontology 47,* 1108–1112.

Han, T.M. & Runnegar, B. 1992: Megascopic eukaryotic algae from the 2.1-billion-year-old Negaunee Iron-Formation, Michigan. *Science 257,* 232–235.

Hanic, L.A. & Craigie, J.S. 1969: Studies on the algal cuticle. *Journal of Phycology 5,* 89–102.

Harris, T.M. 1974: *Williamsoniella lignieri*: its pollen and the compression of spherical pollen grains. *Palaeontology 17,* 125–148.

Hatcher, P.G., Spiker, E.C., Szeverenyi, N.M. & Maciel, G.E. 1983: Selective preservation and origin of petroleum-forming aquatic kerogen. *Nature 305,* 498–501.

Haxo, F.T., Lewin, R.A., Lee, K.W. & Li, M. 1987: Fine structure and pigments of *Oscillatoria* (*Trichodesmium*) aff. *thiebautii* (Cyanophyta) in culture. *Phycologia 26,* 443–456.

Heaman, L.M., LeCheminant, A.N. & Rainbird, R.H. 1992: Nature and timing of Franklin igneous events, Canada: implications for a Late Proterozoic mantle plume and the break-up of Laurentia. *Earth and Planetary Science Letters 109,* 117–131.

Hendriks, L., De Baere, R., Van de Peer, Y., Neefs, J., Goris, A. & De Wachter, R. 1991: The evolutionary position of the rhodophyte *Porphyra umbilicalis* and the basidiomycete *Leucosporidium scottii* among other eukaryotes as deduced from complete sequences of small ribosomal subunit RNA. *Journal of Molecular Evolution 32,* 167–177.

Hermann, T.N. 1974: Nakhodki massovykh skoplenij trikhomov v rifee. [Finds of massive accumulations of trichomes in the Riphean.] *In* Timofeev, B.V. (ed.): *Mikrofitofossilii Proterozoia i Rannego Paleozoia SSSR,* 6–10. Nauka, Leningrad.

Hermann, T.N. 1979: Nakhodki gribov v rifee. [Finds of fungi in the Riphean.] *In* Sokolov, B.S. (ed.): *Paleontologiya Dokembriya i Rannego Kembriya.* 129–136, Nauka, Leningrad.

Hermann, T.N. 1981a: Nakhodki nitchatykh vodoroslej v Miroedikhinskoj svite verkhnego dokembriya. [Filamentous algae from the Miroyedikha Formation of the upper Precambrian. *Paleontological Journal 1981:4,* 111–116.] *Paleontologicheskij Zhurnal 1981:4,* 118–121.

Hermann, T.N. 1981b: Nitchatye mikroorganizmy Lakhandinskoj svity reki Mai. [Filamentous microorganisms in the Lakhanda Formation on the Maya River. *Paleontological Journal 1981:2,* 100–107.] *Paleontologicheskij Zhurnal 1981:2,* 126–131.

Hermann, T.N. 1986: O nakhodkakh sinezelenykh nitchatykh vodoroslej v verkhnem dokembrii (Miroedikhinskoj svite). [On finds of filamentous blue-green algae in the upper Precambrian (Miroedikha Suite).] *In* Sokolov, B.S. (ed.): *Aktual'nye Voprosy Sovremennoj Paleoalgologii,* 37–40. Naukova Dumka, Kiev.

Hermann, T.N. & Timofeev, B.V. 1985: *Eosolenides* – novaya gruppa problematicheskikh organizmov pozdnego dokembriya. [*Eosolenides* – a new group of problematic organisms from the late Precambrian.] *In*: *Problematiki Pozdnego Dokembriya i Paleozoya.* 9–15. *Trudy Instituta Geologii i Geofiziki SO AN SSSR 632.* Nauka, Moscow.

Hoek, C., van den 1984: The systematics of the Cladophorales. *In* Irvine, D.E.G. & John, D.M. (eds.): *Systematics of the Green Algae.* 157–178. Academic Press, London.

Hofmann, H.J. 1976: Precambrian microflora, Belcher Islands, Canada: significance and systematics. *Journal of Paleontology 50,* 1040–1073.

Hofmann, H.J. 1977: The problematic fossil *Chuaria* from the late Precambrian Uinta Mountain Group, Utah. *Precambrian Research 4,* 1–11.

Hofmann, H.J. 1985a: The mid-Proterozoic Little Dal macrobiota, Mackenzie Mountains, north-west Canada. *Palaeontology 28,* 331–354.

Hofmann, H.J. 1985b: Precambrian carbonaceous megafossils. *In* Toomey, D.F. & Nitecki, M.H. (eds.): *Paleoalgology: Contemporary Research and Applications,* 20–33. Springer-Verlag, Berlin.

Hofmann, H.J. 1992. Proterozoic and selected Cambrian megascopic carbonaceous films. *In* Schopf, J.W. & Klein, C. (eds.): *The Proterozoic Biosphere: A Multidisciplinary Study.* 957–979. Cambridge University Press, Cambridge.

Hofmann, H.J. & Aitken, J.D. 1979: Precambrian biota from the Little Dal Group, Mackenzie Mountains, northwestern Canada. *Canadian Journal of Earth Sciences 16,* 150–166.

Hofmann, H.J. & Chen, J. 1981: Carbonaceous megafossils from the Precambrian (1800 Ma) near Jixian, northern China. *Canadian Journal of Earth Sciences 18,* 443–447.

Hofmann, H.J. & Jackson, G.D. 1991: Shelf-facies microfossils from the Uluksan Group (Proterozoic Bylot Supergroup), Baffin Island, Canada. *Journal of Paleontology 65,* 361–382.

Horodyski, R.J. 1980: Middle Proterozoic shale-facies microbiota from the lower Belt Supergroup, Little Belt Mountains, Montana. *Journal of Paleontology 54,* 649–663.

Horodyski, R.J. 1982: Problematic bedding-plane markings from the Middle Proterozoic Appekunny Argillite, Belt Supergroup, northwestern Montana. *Journal of Paleontology 56,* 882–889.

Horodyski, R.J. & Bloeser, B. 1983: Possible eukaryotic algal filaments from the late Proterozoic Chuar Group, Grand Canyon, Arizona. *Journal of Paleontology 57,* 321–326.

Horodyski, R.J. & Donaldson, J.A. 1980: Microfossils from the Middle Proterozoic Dismal Lakes Group, Arctic Canada. *Precambrian Research 11,* 125–159.

Horodyski, R.J. & Mankiewicz, C. 1990: Possible Late Proterozoic skeletal algae from the Pahrump Group, Kingston Range, southeastern California. *American Journal of Science 290-A,* 149–169.

Horodyski, R.J., Bloeser, B. & Vonder Haar, S. 1977: Laminated algal mats from a coastal lagoon, Laguna Mormorna, Baja California, Mexico. *Journal of Sedimentary Petrology 47,* 680–696.

Hu Yunxu, Jian Wanchou & Hua Hong 1993: Shelly fossils from Middle Proterozoic in Luonan, Shaanxi Province. *Northwest Geoscience 14,* 97–106.

Jankauskas, T.V. 1978: Rastitelnye mikrofossilii iz rifejskikh otlozhenij Yuzhnogo Urala. [Vegetable microfossils from Riphean deposits of the Southern Urals.] *Doklady Akademii Nauk SSSR 242,* 913–915.

Jankauskas, T.V. 1980a: Shishenyakskaya mikrobiota verkhnego rifeya Yuzhnogo Urala. [Shishenyak microbiota from the upper Riphean of the Southern Urals.] *Doklady Akademii Nauk SSSR 251,* 190–192.

Jankauskas, T.V. 1980b: Novye vodorosli iz verkhnego rifeya Yuzhnogo Urala i Priural'ya. [New algae from the Upper Riphean of the Southern Urals and the Ural forelands. *Paleontological Journal 1980:4,* 113–121.] *Paleontologicheskij Zhurnal 1980:4,* 107–113.

Jankauskas, T.V., Mikhailova, N.S. & Hermann, T.N. 1987: V vsesoyuznyj kollokvium po mikrofossiliyam dokembriya SSSR [The V All-Union conference on the Precambrian microfossils of the USSR]. *Izvestiya Akademii Nauk SSSR, Seriya Geologicheskaya 1987:9,* 137–139.

Jankauskas, T.V., Mikhailova, N.S. & Hermann, T.N. (eds.) 1989: *Mikrofossilii Dokembriya SSSR.* [*Precambrian Microfossils of the USSR.*] 190 pp. Nauka, Leningrad.

Jenkins, R.J.F., McKirdy, D.M., Foster, C.B., O'Leary, T. & Pell, S.D. 1992: The record and stratigraphic implications of organic-walled microfossils from the Ediacaran (terminal Proterozoic) of South Australia. *Geological Magazine 129,* 401–410.

Jewell, W.J. & McCarty, P.L. 1971: Aerobic decomposition of algae. *Environmental Science and Technology 5,* 1023–1031.

Jux, U. 1977: Über die Wandstrukturen sphaeromorpher Acritarchen: *Tasmanites* Newton, *Tapajonites* Sommer & Van Boekel, *Chuaria* Walcott. *Palaeontographica B 160,* 1–16.

Kaplan, D.R. & Hagemann, W. 1991: The relationship of cell and organism in vascular plants. *BioScience 41,* 693–703.

Karl, D.M., Knauer, G.A. & Martin, J.H. 1988: Downward flux of particulate organic matter in the ocean: a particle decomposition paradox. *Nature 332,* 438–441.

Kauffman, S.A. 1987: Developmental logic and its evolution. *BioEssays 6,* 82–87.

Kaufman, A.J., Knoll, A.H. & Awramik, S.M. 1992: Biostratigraphic and chemostratigraphic correlation of Neoproterozoic sedimentary successions: Upper Tindir Group, northwestern Canada, as a test case. *Geology 20,* 181–185.

Kidwell, S.M. & Baumiller, T. 1990: Experimental disintigration of regular echinoids: roles of temperature, oxygen, and decay thresholds. *Paleobiology 16,* 247–271.

Kiebar, R.J., Zhou Xianlinag & Mopper, K. 1990: Formation of carbonyl compounds from UV-induced photodegradation of humic substances in natural waters: Fate of riverine carbon in the sea. *Limnology and Oceanology 35,* 1503–1515.

Kirchner, O. 1898: Schizophyceae. *In* Engler, A. & Prantl, K. (eds.): *Die natürlichen Pflanzenfamilien, Volume I, 1a,* 4–92.

Klein, C., Beukes, N.J. & Schopf, J.W. 1987: Filamentous microfossils in the Early Proterozoic Transvaal Supergroup: their morphology, significance, and paleoenvironmental setting. *Precambrian Research 36,* 81–94.

Knoll, A.H. 1981: Paleoecology of late Precambrian microbial assemblages. *In* Niklas, K.J. (ed.): *Paleobotany, Paleoecology and Evolution, Volume 1.* 17–54. Praeger, New York.

Knoll, A.H. 1982: Microfossils from the late Precambrian Draken Conglomerate, Ny Friesland, Svalbard. *Journal of Paleontology 56,* 755–790.

Knoll, A.H. 1984: Microbiotas of the late Precambrian Hunnberg Formation, Nordaustlandet, Svalbard. *Journal of Paleontology 58,* 131–162.

Knoll, A.H. 1985: Exceptional preservation of photosynthetic organisms in silicified carbonates and silicified peats. *Philosophical Transactions of the Royal Society of London B 311,* 111–122.

Knoll, A.H. 1991: End of the Proterozoic Eon. *Scientific American 265,* 64–73.

Knoll, A.H. 1992: Vendian microfossils in metasedimentary cherts of the Scotia Group, Prins Karls Forland, Svalbard. *Palaeontology 35,* 751–774.

Knoll, A.H. & Butterfield, N.J. 1989: New window on Proterozoic life. *Nature 337,* 602–603.

Knoll, A.H. & Calder, S. 1983: Microbiotas of the late Precambrian Ryssö Formation, Nordaustlandet, Svalbard. *Palaeontology 26,* 467–496.

Knoll, A.H. & Golubic, S. 1979: Anatomy and taphonomy of a Precambrian algal stromatolite. *Precambrian Research 10,* 115–151.

Knoll, A.H. & Swett, K. 1990: Carbonate deposition during the Late Proterozoic era: an example from Spitsbergen. *American Journal of Science 290-A,* 104–132.

Knoll, A.H., Hayes, J.M., Kaufman, A.J., Swett, K. & Lambert, I.B. 1986: Secular variation in carbon isotope ratios from Upper Proterozoic successions of Svalbard and East Greenland. *Nature 321,* 1–7.

Knoll, A.H., Strother, P.K. & Rossi, S. 1988: Distribution and diagenesis of microfossils from the lower Proterozoic Duck Creek Dolomite, Western Australia. *Precambrian Research 38,* 257–279.

Knoll, A.H., Swett, K. & Burkhardt, E. 1989: Paleoenvironmental distribution of microfossils and stromatolites in the upper Proterozoic Blacklundtoppen Formation, Spitsbergen. *Journal of Paleontology 63,* 129–145.

Knoll, A.H., Swett, K. & Mark, J. 1991: Paleobiology of a Neoproterozoic tidal flat/lagoonal complex: the Draken Conglomerate Formation, Spitsbergen. *Journal of Paleontology 65,* 531–570.

Kolosov, P.N. 1982: *Verkhnedokembrijskie Paleoalgologicheskie Ostatki Sibirskoj Platformy.* [*Upper Precambrian Paleoalgological Residues from the Siberian Platform.*] 94 pp. Nauka, Moscow.

Kolosov, P.N. 1984: *Pozdnedokembrijskie Mikroorganizmy Vostoka Sibirskoj Platformy* [*Late Precambrian Microorganisms from the Eastern Siberian Platform*]. 84 pp. Yakutskij Filial AN SSSR, Yakutsk.

Kuc, M. 1972: Fossil statoblasts of *Cristatella mucedo* Cuvier in the Beaufort Formation and in interglacial and postglacial deposits of the Canadian Arctic. *Geological Survey of Canada, Paper 72-28,* 12 pp.

Kützing, F.T. 1843: *Phycologia Generalis.* 458 pp. F. A. Brockhaus, Leipzig.

Kützing, F.T. 1845: *Phycologia Germanica.* 340 pp. W. Köhne, Nordhausen.

Lanier, W.P. 1988: Structure and morphogenesis of microstromatolites from the Transvaal Supergroup, South Africa. *Journal of Sedimentary Petrology 58,* 89–99.

Leo, R.F. & Barghoorn, E.S. 1976: Silicification of wood. *Harvard University Botanical Museum Leaflets 25*, 47 pp.

Licari, G.R. 1978: Biogeology of the late pre-Phanerozoic Beck Spring Dolomite of eastern California. *Journal of Paleontology 52*, 767–792.

Liu Chili 1982: Micro-fossil algal communities from the Wumishan Formation in Jixian, China, and their geological significance. *Bulletin of Nanjing University - Special Paper on Algae*, 121–166.

Liu Xue-xian, Liu Zhili, Zhang Lun & Xu Xuesi 1984: A study of late Precambrian microfossil algal community from Suining County, Jiangsu Province. *Acta Micropalaeontologica Sinica 1*, 171–182.

Liu Zhili & Li Hanming 1986: Fossil blue-green algal community from the upper Permian of Guangxi and its significant role in forming bottom of coal beds. *Acta Micropalaeontologica Sinica 3*, 261–272.

Lo, S.C. 1980: Microbial fossils from the lower Yudoma Suite, earliest Phanerozoic, eastern Siberia. *Precambrian Research 13*, 109–166.

Lochte, K. & Turley, C.M. 1988: Bacteria and cyanobacteria associated with phytodetritus in the deep sea. *Nature 333*, 67–69.

Lu Xinglin & Zhu Liying. 1985: Ultramicrofossils and microfossils from Early Palaeozoic meta-boghead coals and carbonolites in western Hunan. *In: Selected Papers from the 1th National Fossil Algal Symposium*. 83–88. Geological Publishing House, Beijing.

Luo Qiling 1985: Micropalaeoflora from upper Precambrian Shijia Formation in northern Anhui Province and its stratigraphic significance. *Bulletin of the Tianjin Institute, Geology and Mineral Resources 12*, 169–182.

Luo Qiling, Wang Fuxing & Wang Yangeng 1982: Uppermost Sinian – lowerest Cambrian age microfossils from Qingzhenzhijin County, Guizhou Province. *Bulletin of the Tianjin Institute, Geology and Mineral Resources 6*, 23–41.

Mädler, K. 1964: Bemerkenswerte Sporenformen aus dem keuper und unteren Lias. *Fortschritte in der Geologie von Rheinland und Westfalen 12*, 169–200.

Maithy, P.K. 1975: Micro-organisms from the Bushimay System (Late Pre-Cambrian) of Kanshi, Zaire. *The Palaeobotanist 22*, 133–149.

Maithy, P.K. & Shukla, M. 1977: Microbiota from the Suket Shales, Ramapura, Vindhyan System (Late Pre-Cambrian), Madhya Pradesh. *The Palaeobotanist 23*, 176–188.

Maithy, P.K. & Shukla, M. 1984: Reappraisal of *Fermoria* and allied remains from the Suket Shale Formation, Ramapura. *The Palaeobotanist 32*, 146–152.

Maliva, R.G., Knoll, A.H. & Siever, R. 1990: Secular change in chert distribution: a reflection of evolving biological participation in the silica cycle. *Palaios 4*, 519–532.

Mankiewicz, C. 1992: *Obruchevella* and other microfossils in the Burgess Shale: preservation and affinity. *Journal of Paleontology 66*, 717–729.

Marchant, H.J. 1977: Cell division and colony formation in the green alga *Coelastrum* (Chlorococcales). *Journal of Phycology 13*, 102–110.

Mathur, S.M. 1983: A new collection of fossils from the Precambrian Vindhyan Supergroup of central India. *Current Science 52*, 361–363.

Matsuda, Y. 1988: The *Chlamydomonas* cell walls and their degrading enzymes. *Japanese Journal of Phycology 36*, 246–264.

Mattox, K.R. & Stewart, K.D. 1984: Classification of the green algae. *In* Irvine, D.E.G. & John, D.M. (eds.): *Systematics of the Green Algae*. 29–72. Academic Press, London.

McMenamin, D.S., Kumar, S. & Awramik, S.M. 1983: Microbial fossils from the Kheinjua Formation, Middle Proterozoic Semri Group (Lower Vindhyan), Son Valley area, central India. *Precambrian Research 21*, 247–271.

Mendelson, C.V. & Schopf, J.W. 1982: Proterozoic microfossils from the Sukhaya Tunguska, Shorikha, and Yudoma formations of the Siberian Platform, USSR. *Journal of Paleontology 56*, 42–83.

Mikhailova, N.S. 1986: Novye nakhodki mikrofitofossilij iz otlozhenij verkhnego rifeya Krasnoyarskogo kraya. [New finds of microphytofossils from upper Riphean deposits of the Krasnoyar region.] *In* Sokolov, B.S. (ed.): *Aktual'nye Voprosy Sovremennoj Paleoalgologii*. 31–37. Naukova Dumka, Kiev.

Muir, M.D. 1976: Proterozoic microfossils from the Amelia Dolomite, McArthur Basin, Northern Territory. *Alcheringa 1*, 143–158.

Nägeli, C. 1849: *Gattungen Einzelliger Algen, Physiologisch und Systematisch Bearbeitet*. 139 pp. F. Schulthess, Zürich.

Naumova, S.N. 1949: Spory nizhnego kembriya. [Spores from the Lower Cambrian.] *Izvestiya Akademii Nauk SSSR, Seriya Geologicheskaya 1949:4*, 49–56.

Nautiyal, A.C. 1980: Cyanophycean algal remains and paleoecology of the Precambrian Gangolihat Dolomites Formation of the Kumaun Himalaya. *Indian Journal of Earth Sciences 7*, 1–11.

Nautiyal, A.C. 1982: Algal remains from the Random Formation (Late Precambrian) of Newfoundland, Canada. *Indian Journal of Earth Sciences 9*, 174–177.

Nyberg, A.V. & Schopf, J.W. 1984: Microfossils in stromatolitic cherts from the upper Proterozoic Min'yar Formation, Southern Ural Mountains, USSR. *Journal of Paleontology 58*, 738–772.

Oehler, J.H. 1977: Microflora of the H.Y.C. Pyritic Shale Member of the Barney Creek Formation (McArthur Group), middle Proterozoic of northern Australia. *Alcheringa 1*, 315–349.

Ogurtsova, R.N. & Sergeev, V.N. 1987: Mikrobiota Chichkanskoj svity verkhnego dokembriya Malogo Karatau (yuzhnyj Kazakhstan). [The microbiota of the upper Precambrian Chichkan Formation in the Lesser Karatau region (southern Kazakhstan). *Paleontological Journal 1987:2*, 101–112.] *Paleontologicheskij Zhurnal 1987:2*, 107–116.

Ogurtsova, R.N. & Sergeev, V.N. 1989: Megasferomorfidy Chichkanskoj svity verkhnego dokembriya yuzhnogo Kazakhstana. [Megaspheromorphs from the upper Precambrian Chichkanskaya Formation in southern Kazakhstan. *Paleontological Journal 1989:2*, 117–121.] *Paleontologicheskij Zhurnal 1989:2*, 119–122.

O'Kelly, C.J. & Floyd, G.L. 1984: Correlations among patterns of sporangial structure and development, life histories, and ultrastructural features in the Ulvophyceae. *In* Irvine, D.E.G. & John, D.M. (eds.): *Systematics of the Green Algae*, 121–156. Academic Press, London.

Oltmanns, F. 1904: *Morphologie und Biologie der Algen, Volume 1*. 733 pp. G. Fischer, Jena.

Ouyang Shu, Yin Leiming & Li Zaiping 1974: Sinian and Cambrian spores. *In: A Handbook of the Stratigraphy and Paleontology of Southwestern China*, 72–80, 114–123. Science Publishing House, Beijing.

Pascher, A. 1914: Über Flagellaten und Algen. *Berichte der Deutschen Botanischen Gesellschaft 32*, 136–160.

Perasso, R., Baroin, A., Qu, L.H., Bachellerie, J.P. & Adoutte, A. 1989: Origin of the algae. *Nature 339*, 142–144.

Philp, R.P. & Calvin, M. 1976: Possible origin for insoluble organic (kerogen) debris in sediments from insoluble cell-wall materials of algae and bacteria. *Nature 262*, 134–136.

Pjatiletov, V.G. 1979: O nakhodkakh sinezelenykh vodoroslej v Yudomskikh otlozheniyakh Yakutii (vend). [On finds of blue-green algae in Yudoma deposits of Yakutia (Vendian).] *Doklady Akademii Nauk SSSR 249*, 714–715.

Pjatiletov, V.G. 1980: Yudomskij kompleks mikrofossilij yuzhnoj Yakutii. [Yudoma complex microfossils from southern Yakutia.] *Geologiya i Geofizika 21*, 8–20.

Pjatiletov, V.G. 1988: Mikrofitofossilii pozdnego dokembriya Uchuro-Majskogo rajona. [Microphytofossils from the late Precambrian of the Uchur-Maia region.] *In* Khomentovsky, V.V. & Schenfil, V.Y. (eds.): *Pozdnij Dokembrij i Rannij Paleozoj Sibiri Rifej i Vend*, 47–104. Institut Geologii i Geofiziki SO AN SSSR, Novosibirsk.

Pjatiletov, V.G. & Karlova, G.A. 1980: Verkhnerifejskij kompleks rastitel'nykh mikrofossilii Yenisejskogo kryazha. [Upper Riphean complex plant microfossils from the Yenisey region.] *In* Khomentovsky, V.V. (ed.): *Novye Dannye po Stratigrafii Pozdnego Dokembriya Zapada Sibirskoj Platformy i ee Skladchatogo Obramleniya*, 56–71. Institut Geologii i Geofiziki SO AN SSSR, Novosibirsk.

Puel, F., Largeau, C. & Giraud, G. 1987: Occurrence of a resistant biopolymer in the outer walls of the parasitic alga *Prototheca wickerhamii* (Chlorococcales): ultrastructural and chemical studies. *Journal of Phycology 23*, 649–656.

Rabenhorst, L. 1863: *Kryptogamen-Flora von Sachsen, der Ober-Lausitz, Thüringen und Nordböhmen, mit Berüchsichtigung der Benachbarten Länder*. 653 pp. Eduard Kummer, Leipzig.

Reitlinger, E.A. 1948: Kembrijskie foraminifery Yakutii. [Cambrian fora-minifera of Yakutia.] *Biulleten' Moskovskogo Obshchestva Ispytatelej Prirody (Novaya Seriya) 23*, 77–84.

Revsbech, N.P., Jorgensen, B.B. & Blackburn, T.H. 1980: Oxygen in the sea bottom measured with a microelectrode. *Science 207*, 1355–1356.

Rex, G. 1983: The compression state of preservation of Carboniferous lepidodendrid leaves. *Review of Palaeobotany and Palynology 39*, 65–85.

Rex, G.M. & Chaloner, W.G. 1983: The experimental formation of plant compression fossils. *Palaeontology 26*, 231–252.

Round, F.E. 1963: The taxonomy of the Chlorophyta. *British Phycological Bulletin 2*, 224–235.

Rowell, A.J. 1971: Supposed pre-Cambrian brachiopods. *Smithsonian Contributions to Paleobiology 3*, 71–79.

Runnegar, B. 1982: A molecular-clock date for the origin of the animal phyla. *Lethaia 15*, 199–205.

Schenfil, V.Yu. 1980: Obruchevelly v rifejskikh otlozheniyakh Yenisej-skogo kryazha. [*Obruchevella* representatives in the Riphean deposits of the Yenisey ridge region.] *Doklady Akademii Nauk SSSR 254*, 993–994.

Schenfil, V.Yu. 1983: Vodorosli v dokembrijskikh otlozheniyakh vostoch-noj Sibiri. [Algae in the Precambrian sediments of eastern Siberia.] *Doklady Akademii Nauk SSSR 269*, 471–473.

Schepeleva, E.D. 1960: Nakhodki sinezelenykh vodoroslej v nizhnekem-brijskikh otlozheniyakh Leningradskoj oblasti. [Finds of blue-green algae in Lower Cambrian deposits of the Leningrad region.] *In: Prob-lemy Neftyanoj Geologii i Voprosy Metodiki Laboratornykh Issledovanij.* 170–172. Nauka, Moscow.

Schopf, J.W. 1968: Microflora of the Bitter Springs Formation, late Pre-cambrian, central Australia. *Journal of Paleontology 42*, 651–688.

Schopf, J.W. 1992: Proterozoic prokaryotes: affinities, geologic distribu-tion, and evolutionary trends. *In* Schopf, J.W. & Klein, C. (eds.): *The Proterozoic Biosphere: A Multidisciplinary Study.* 195–218. Cambridge University Press, Cambridge.

Schopf, J.W. & Barghoorn, E.S. 1969: Microorganisms from the late Pre-cambrian of South Australia. *Journal of Paleontology 43*, 111–118.

Schopf, J.W. & Blacic, J.M. 1971: New microorganisms from the Bitter Springs Formation (late Precambrian) of the north-central Amadeus Basin, Australia. *Journal of Paleontology 45*, 925–960.

Sears, J.R. & Brawley, S.H. 1982: *Smithsoniella* gen. nov., a possible evolu-tionary link between the multicellular and siphonous habits in the Ulvophyceae, Chlorophyta. *American Journal of Botany 69*, 1450–1461.

Seilacher, A., Reif, W.-E. & Westphal, F. 1985: Sedimentological, ecological and temporal patterns of fossil Lagerstätten. *Philosophical Transactions of the Royal Society of London B 311*, 5–23.

Sergeev, V.N. 1991: Okremnennye mikrofossilij dokembriya Urala i Kazakhstana i ikh biostratigraficheskie vozmozhnosti. [Silicified micro-fossils from the Precambrian of the Urals and Kazakhstan and their biostratigraphic potential.] *Izvestya Akademii Nauk SSSR, Seriya Geo-logicheskaya 1991:11*, 87–97.

Sergeev, V.N. & Krylov, I.N. 1986: Mikrofossilii Min'yarskoj svity Urala. [Microfossils of the Min'yar Formation of the Urals. *Paleontological Journal 1986:1*, 63–75.] *Paleontologicheskij Zhurnal 1986:1*, 84–95.

Sharp, L.W. 1934: *Introduction to Cytology, 3rd Ed..* 567 pp. McGraw-Hill, New York.

Shukovsky, E.S. & Halfen, L.N. 1976: Cellular differentiation of terminal regions of trichomes of *Oscillatoria princeps* (Cyanophyceae). *Journal of Phycology 12*, 336–342.

Slack, J.M.W., Holland, P.W.H. & Graham, C.F. 1993: The zootype and the phylotypic stage. *Nature 361*, 490–492.

Sogin, M.L. 1989: Phylogenetic meaning of the Kingdom concept: an unusual ribosomal RNA from *Giardia lamblia*. *Science 243*, 75–77.

Sokolov, B.S. & Ivanovskij, A.B. (eds.) 1985: *Vendskaya Sistema [Vendian System]*, Volume 1. 221 pp. Nauka, Moscow.

Song Xueliang 1982: Microfossils and Acritarchs. *In* Luo Huilin *et al.*: *The Sinian–Cambrian Boundary in Eastern Yunnan, China.* 216–222. Yun-nan Institute of Geological Sciences.

Sovetov, Yu.K. & Schenfil, V.Yu. 1977: Novaya dokembrijskaya mikro-biota (yuzhnyj Kazakhstan). [New Precambrian microbiota from

southern Kazakhstan.] *Doklady Akademii Nauk SSSR 232*, 1193–1196.

Stanier, R.Y., Sistrom, W.R., Hansen, T.A., Whitton, B.A., Castenholz, R.W., Pfenning, N., Gorlenko, V.N., Kondratieva, E.N., Eimhjellen, K.E., Whittenbury, R., Gherna, R.L. & Trüper, H.G. 1978: Proposal to place nomenclature of the Cyanobacteria (blue-green algae) under the rules of the International Code of Nomenclature of Bacteria. *Interna-tional Journal of Systematic Bacteriology 28*, 335–336.

Staplin, F.L., Jansonius, J. & Pocock, S.A.J. 1965: Evaluation of some acritarchous hystrichosphere genera. *Neues Jahrbuch für Geologie und Paläontologie, Abhandlungen 123*, 167–201.

Strother, P.K., Knoll, A.H. & Barghoorn, E.S. 1983: Micro-organisms from the late Precambrian Narssârssuk Formation, north-western Green-land. *Palaeontology 26*, 1–32.

Sun Weiguo 1987a: Palaeontology and biostratigraphy of late Precambrian macroscopic colonial algae: *Chuaria* Walcott and *Tawuia* Hofmann. *Palaeontographica B 203*, 109–134.

Sun Weiguo 1987b: Discussions on the age of the Liulaobei Formation. *Precambrian Research 36*, 349–352.

Tandon, K.K. & Kumar, S. 1977: Discovery of annelid and arthropod remains from lower Vindhyan rocks (Precambrian) of central India. *Geophytology 7*, 126–129.

Tappan, H. 1980: *The Paleobiology of Plant Protists.* 1028 pp. W.H. Free-man, San Francisco.

Tegelaar, E.W., de Leeuw, J.W., Derenne, S. & Largeau, C. 1989: A reap-praisal of kerogen formation. *Geochimica et Cosmochimica Acta 53*, 3103–3106.

Tehler, A. 1988: A cladistic outline of the Eumycota. *Cladistics 4*, 227–277.

Theng, B.K.G. 1979: *Formation and Properties of Clay-Polymer Complexes.* 362 pp. Elsevier, Amsterdam.

Thuret, G. 1875: Essai de classification de Nostochinées. *Annales Sciences Naturelle (Botanique) 6*, 372–382.

Timofeev, B.V. 1966: *Mikropaleofitologicheskoe Issledovanie Drevnikh Svit.* [*Micropaleophytological Investigations of Ancient Suites.*] 147 pp. Nauka, Moscow.

Timofeev, B.V. 1969: *Sferomorfidy Proterozoya.* [*Proterozoic Spheromor-phida.*] 145 pp. Nauka, Leningrad.

Timoféev, B.V. 1970: Sphaeromorphida géants dans le Précambrien avan-cé. *Review of Palaeobotany and Palynology 10*, 157–160.

Timofeev, B.V. & Hermann, T.N. 1979: Dokembrijskaya mikrobiota La-khandinskoj svity. [Precambrian microbiota of the Lakhandin Suite.] *In* Sokolov, B.S. (ed.): *Paleontologiya Dokembriya i Rannego Kembriya.* 137–147. Nauka, Leningrad.

Timofeev, B.V., Hermann, T.N. & Mikhailova, N.S. 1976: *Microfitofossilii Dokembriya, Kembriya i Ordovika.* [*Microphytofossils of the Precam-brian, Cambrian and Ordovician.*] 106 pp. Nauka, Leningrad.

Tynni, R. & Donner, J. 1980: A microfossil and sedimentation study of the Late Precambrian formation of Hailuoto, Finland. *Geological Survey of Finland, Bulletin 311*, 27 pp.

Valensi, L. 1949: Sur quelques microorganismes planctoniques des silex du Jurassique moyen du Poitou et de Normandie. *Bulletin de la Société Géologique de France, 5e série 18*, 537–550.

Vidal, G. 1976: Late Precambrian microfossils from the Visingsö Beds in southern Sweden. *Fossils and Strata 9*, 57 pp.

Vidal, G. 1990: Giant acanthomorph acritarchs from the upper Proterozoic in southern Norway. *Palaeontology 33*, 287–298.

Vidal, G. & Ford, T.D. 1985: Microbiotas from the late Proterozoic Chuar Group (northeastern Arizona) and Uinta Mountain Group (Utah) and their chronostratigraphic implications. *Precambrian Research 28*, 349–389.

Vidal, G., Moczydłowska, M. & Rudavskaya, V.A. 1993: Biostratigraphical implications of a *Chuaria–Tawuia* assemblage and associated acritarchs from the Neoproterozoic of Yakutia. *Palaeontology 36*, 387–402.

Walcott, C.D. 1899: Pre-Cambrian fossiliferous formations. *Geological Society of America, Bulletin 10*, 199–244.

Walcott, C.D. 1919: Cambrian geology and paleontology IV. No. 5 – Middle Cambrian algae. *Smithsonian Miscellaneous Collections 67*, 217–260.

Walter, M.R., Oehler, J.H. & Oehler, D.Z. 1976: Megascopic algae 1300 million years old from the Belt Supergroup, Montana: a reinterpretation of Walcott's *Helminthoidichnites*. *Journal of Paleontology 50*, 872–881.

Walter, M.R., Du Rulin & Horodyski, R.J. 1990: Coiled carbonaceous megafossils from the Middle Proterozoic of Jixian (Tianjin) and Montana. *American Journal of Science 290-A*, 133–148.

Wang Fuxing, Zhang Xuanyang & Guo Ruihuan 1983: The Sinian microfossils from Jinning, Yunnan, south west China. *Precambrian Research 23*, 133–175.

Wang Guixiang 1982: Late Precambrian Annelida and Pogonophora from the Huainan of Anhui Province. *Bulletin of the Tianjin Institute, Geology and Mineral Resources 6*, 9–22.

Weiss, A.F. 1984: Mikrofossilii iz verkhnego rifeya Turukhanskogo rajona. [Microfossils from the Upper Riphean of the Turukhansk region. *Paleontological Journal 1984:2*, 98–104.] *Paleontologicheskij Zhurnal 1984:2*, 102–108.

Wettstein, R. 1924: *Handbuch der Systematischen Botanik, Volume 1.* 467 pp. Franz Deuticke, Leipzig.

Wiebe, W.J., Sheldon, W.M.J. & Pomeroy, L.R. 1992: Bacterial growth in the cold: evidence for an enhanced substrate requirement. *Applied and Environmental Microbiology 58*, 359–364.

Wilson, C.B. 1961: The upper middle Hecla Hoek rocks of Ny Friesland, Spitsbergen. *Geological Magazine 98*, 89–116.

Wiman, C. 1894: Ein präkambrisches Fossil. *Bulletin of the Geological Institution of the University of Upsala 2*, 109–113.

Woese, C.R. & Fox, G.E. 1977: Phylogenetic structure of the prokaryotic domain: The primary kingdoms. *Proceedings of the National Academy of Sciences USA 74*, 5088–5090.

Woese, C.R., Kandler, O. & Wheelis, M.L. 1990: Towards a natural system of organisms: Proposal for the domains Archaea, Bacteria, and Eucarya. *Proceedings of the National Academy of Sciences USA 87*, 4576–4579.

Xing Yusheng, Duan Chenghua, Liang Yuzuo, Cao Renguan, *et al.* 1985: *Late Precambrian Paleontology of China. People's Republic of China Ministry of Geology and Mineral Resources, Geological Memoirs, Series 2, Number 2*, 243 pp.

Xu Zhao-liang 1984: Investigation on the procaryotic microfossils from the Gaoyuzhuang Formation, Jixian, North China. *Acta Botanica Sinica 26*, 216–222, 312–319.

Yakschin, M.S. 1991: *Vodoroslevaya mikrobiota nizhnego rifeya Anabarskogo podnyatiya (Kotujkanskaya svita).* [*Algal Microbiota from the Early Riphean of the Anabar Uplift (Kotujkan Suite).*] 64 pp. *Trudy Instituta Geologii i Geofiziki SO AN SSSR 768*. Nauka, Novosibirsk.

Yakshin, M.S. & Luchinina, V.A. 1981: Novye dannye po iskopaemym vodoroslyam semejstva Oscillatoriaceae (Kirchn.) Elenkin. [New data on fossil algae of the family Oscillatoriaceae (Kirchn.) Elenkin.] *In* Meshkova, N.P. & Nukolaeva, I.V. (eds.): *Pogranichnye otlozheniya dokembriya i kembriya Sibirskoj Platformy (biostratigrafiya, paleontologiya, usloviya obrazovaniya).* 28–34. Nauka, Novosibirsk.

Yan Yushong 1989: Shale-facies algal filaments in Jixian County. *Bulletin of the Tianjin Institute, Geology and Mineral Resources 21*, 149–165.

Yan Yu-shong & Zhu Shi-xing 1992: Discovery of acanthomorphic acritarchs from the Baicaoping Formation in Yongji, Shanxi and its geological significance. *Acta Micropalaeontologica Sinica 9*, 267–282.

Yin Chongyu 1985: Micropalaeoflora from the late Precambrian in Huainan region of Anhui Province and its stratigraphic significance. *Professional Papers of Stratigraphy and Paleontology 12*, 97–119.

Yin Chongyu 1990: Spinose acritarchs from the Toushantuo Formation in the Yangtze Gorges and its geological significance significance. *Acta Micropalaeontologica Sinica 7*, 265–270.

Yin Leiming. 1979: Microflora from the Anshan Group and the Liaohe Group in E. Liaoning with its stratigraphic significance. *In: Selected Works for a Scientific Symposium on Iron-Geology and Palaeontology.* 39–60. Science Press, Beijing.

Yin Leiming 1985a: Microfossils of the Doushantuo Formation in the Yangtze Gorge district, western Hubei. *Palaeontologia Cathayana 2*, 229–249.

Yin Leiming 1985b: A glimpse of Precambrian microfossils. *In: Selected Papers from the 1th National Fossil Algal Symposium.* 179–186. Geological Publishing House, Beijing.

Yin Leiming 1987: Microbiotas of latest Precambrian sequences in China. *In: Stratigraphy and Palaeontology of Systemic Boundaris in China. Precambrian-Cambrian Boundary (1).* 415–494. Nanjing University Publishing House, Nanjing.

Yin Leiming & Li Zaiping 1978: Precambrian microfloras of southwest China, with reference to their stratigraphical significance. *Nanjing Institute of Geology and Palaeontology, Memoir 10*, 41–102.

Zalessky, M.D. 1926: Premières observations microscopiques sur le schiste bitumineux du Volgien inférieur. *Societe Geologique du Nord, Annales 51*, 65–104.

Zang, W. & Walter, M.R. 1992a: Late Proterozoic and Cambrian microfossils and biostratigraphy, Amadeus Basin, central Australia. *Association of Australasian Palaeontologists, Memoir 12*. 132 pp.

Zang Wenlong & Walter, M.R. 1992b: Late Proterozoic and Early Cambrian microfossils and biostratigraphy, northern Anhui and Jiangsu, central-eastern China. *Precambrian Research 57*, 243–323.

Zechman, F.W., Theriot, E.C., Zimmer, E.A. & Chapman, R.L. 1990: Phylogeny of the Ulvophyceae (Chlorophyta): Cladistic analysis of nuclear encoded rRNA sequence data. *Journal of Phycology 26*, 700–710.

Zelibor, J.J., Romankiw, L., Hatcher, P.G. & Colwell, R.R. 1988: Comparative analysis of the chemical composition of mixed and pure cultures of green algae and their decomposed residues by ^{13}C nuclear magnetic resonance spectroscopy. *Applied and Environmental Microbiology 54*, 1051–1060.

Zhang Pengyuan 1981: Microfossil blue-green algae from the Wumishan Formation of Jixian. *Acta Geologica Sinica 55*, 253–257.

Zhang Pengyuan 1982: Microfossils from the Wumishan Formation of Jixian County. *Acta Geologica Sinica 56*, 34–41.

Zhang Pengyuan 1983: Microbiota from chert facies of Wumishan Formation, Jixian County. *Bulletin of the Tianjin Institute, Geology and Mineral Resources 8*, 207–223.

Zhang Pengyuan 1987: Microfossils from the Middle Proterozoic in Kuancheng County, eastern Hebei, China. *Professional Papers of Stratigraphy and Paleontology 17*, 263–276.

Zhang Pengyuan & Gu Shuqin 1986: Microfossils from the Wumishan Formation of the Jixian System in the Ming Tombs section of Beijing, China. *Acta Geologica Sinica 60*, 321–331.

Zhang Pengyuan & Yan Xiaoli 1984: Microfossils from the Gaoyuzhuang Formation in Laishui County, Hebei, China. *Acta Geologica Sinica 58*, 196–204.

Zhang Pengyuan, Zhu Mu & Song Wu 1989: Middle Proterozoic (1200–1400 Ma) microfossils from the Western Hills near Beijing, China. *Canadian Journal of Earth Sciences 26*, 322–328.

Zhang Renjie, Feng Shaonan, Ma Guogan, Xu Guanghong & Yan Daoping 1991: Late Precambrian macroscopic fossil algae from Hainan Island. *Acta Palaeontologica Sinica 30*, 115–125.

Zhang Yun 1981: Proterozoic stromatolite microfloras of the Gaoyuzhuang Formation (early Sinian:Riphean), Hebei, China. *Journal of Paleontology 55*, 485–506.

Zhang Yun 1985: Stromatolitic microbiota from the Middle Proterozoic Wumishan Formation (Jixian Group) of the Ming Tombs, Beijing, China. *Precambrian Research 30*, 277–302.

Zhang Yun 1989: Multicellular thallophytes with differentiated tissues from Late Proterozoic phosphate rocks of South China. *Lethaia 22*, 113–132.

Zhang Yun & Yuan Xun-lai 1992: New data on multicellular thallophytes and fragments of cellular tissues from Late Proterozoic phosphate rocks, South China. *Lethaia 25*, 1–18.

Zhang Zhongying 1981: A new Oscillatoriaceae-like filamentous microfossil from the Sinian (late Precambrian) of western Hubei Province, China. *Geological Magazine 118*, 201–206.

Zhang Zhongying 1982: Upper Proterozoic microfossils from the Summer Isles, N. W. Scotland. *Palaeontology 25*, 443–460.

Zhang Zhongying 1984: A new microphytoplankton species from the Sinian of western Hubei Province. *Acta Botanica Sinica 26*, 94–98.

Zhang Zhongying 1986: New material of filamentous fossil cyanophytes from the Doushantuo Formation (late Sinian) in the eastern Yangtze Gorge. *Scientia Geologica Sinica 1986*, 30–37.

Zheng Wenwu 1980: A new occurrence of fossil group of *Chuaria* from the Sinian System in north Anhui and its geological meaning. *Bulletin of the Chinese Academy of Geological Sciences VI (Tianjin Institute) 1*, 49–69.

Zhu Shixing 1982: A preliminary study of fossil micro-organisms from stromatolites in the lower part of Sinian suberathem, Yanshan Range.

Bulletin of the Tianjin Institute, Geology and Mineral Resources 5, 1–26.

Zhu Shixing, Wang Yangeng & Zang Lin 1984: Formation of the Kaiyang Phosphorites in China as related to ancient microorganisms. *In: IGCP Symposium of 5th International Field Workshop and Seminar on Phosphorite.* 165–193.

Zhu Wei-qing & Chen Meng-e 1984: On the discovery of macrofossil algae from the late Sinian in the eastern Yangtze Gorges, South China. *Acta Botanica Sinica 26*, 558–560.

Appendix

Alphabetic tabulation of all fossil taxa recorded in the Svanbergfjellet Formation and the field samples in which they occur (sample identification is numerically coded and appears in brackets after the name; an asterisk [*] indicates type material). Junior synonyms (as here assessed) are indented and listed alphabetically below each legitimate name. Sample codes correspond to the horizons indicated in Fig. 2: (1) 86-G-63; (2) B-2-2; (3) 86-G-33; (4) 86-G-62; (5) 86-G-61; (6) 86-G-30; (7) 86-G-28; (8) 86-P-89; (9) P-3400; (10) 86-G-22; (11) P-3085; (12) 86-P-82; (13) P-3075; (14) 86-G-9; (15) 86-SP-8; (16) P-2945; (17) SV-3; (18) 86-G-15; (19) 86-G-14; (20) P-2664; (21) P-2628; (22) 86-G-4; (23) 86-G-8.

Brachypleganon khandanum Lo, 1980 [4]

Cephalonyx Weiss, 1984
  Arthrosiphon Weiss, 1984
  Contextuopsis Hermann, 1985
  Rectia Jankauskas, 1989

Cephalonyx geminatus (Jankauskas, 1980) Butterfield, n.comb. *(4)*
  Calyptothrix geminata Jankauskas, 1980

Cerebrosphaera buickii Butterfield, n.gen., n.sp. [3, 4, 5, 6, 7, 16*]

Chlorogloeaopsis zairensis Maithy, 1975 [4]
  Polysphaeroides biseritus Liu, 1985

Chuaria Walcott, 1899
  Fermoria Chapman, 1935
  Luonanconcha Jian & Hu, 1993
  Protobolella Chapman, 1935

Chuaria circularis Walcott, 1899 [3, 4, 5, 7]
  Chuaria annularis Zheng, 1980
  Chuaria minima (Chapman, 1935) Maithy & Shukla, 1984
  Chuaria multirugosa Du, 1985
  Fermoria capsella Chapman, 1935
  Fermoria granulosa Chapman, 1935
  Fermoria minima Chapman, 1935
  Protobolella jonesi Chapman, 1935

Comasphaeridium sp. [4]

Cyanonema sp. [4]

Cymatiosphaeroides kullingii Knoll, 1984 [14, 18, 19, 20, 21, 23]

Dictyotidium fullerene Butterfield, n.sp. [4*]

Digitus adumbratus Butterfield, n.sp. [4]

Eoentophysalis belcherensis Hofmann, 1976 [20, 22]
  Eoentophysalis cumulus Knoll & Golubic, 1979
  Myxococcoides kingii Muir, 1976

Eoentophysalis croxfordii (Muir, 1976) Butterfield, n.comb. [14, 18, 19]
  Ameliaphycus croxfordii Muir, 1976

Eosynechococcus moorei Hofmann, 1976 [20]

Germinosphaera bispinosa Mikhailova, 1986 [4]
  Germinosphaera unispinosa Mikhailova, 1986

Germinosphaera fibrilla (Ouyang, Yin & Li, 1974), Butterfield, n.comb. [4]
  Archaeohystrichosphaeridium truncatum Ouyang, Yin & Li, 1974
  Germinosphaera tadasii Weiss, 1989
  Ooidium fibrillum Ouyang, Yin & Li, 1974

Germinosphaera jankauskasii Butterfield, n.sp. [4*]

Gloeodiniopsis lamellosa Schopf, 1968 [14, 23]

Goniosphaeridium sp. [4]

Gorgonisphaeridium sp. [1]

Leiosphaeridia crassa (Naumova, 1949) Jankauskas, 1989 [1, 2, 3, 4, 5, 6, 7, 16]

Leiosphaeridia jacutica (Timofeev, 1966) Mikhailova & Jankauskas, 1989 [3, 4, 5, 7, 16]

Leiosphaeridia tenuissima Eisenack, 1958 [1, 4, 16]

Leiosphaeridia wimanii (Brotzen, 1941) Butterfield, n.comb. [3, 4, 5, 16]
  Chuaria wimani Brotzen, 1941
  Kildinella magna Timofeev, 1969

Myxococcoides minor Schopf, 1968 [20]

Myxococcoides cantabrigiensis Knoll, 1982 [19]

Obruchevella blandita Schenfil, 1980 [4, 5, 6, 7]
  Glomovertella glomerata (Jankauskas, 1980) Jankauskas, 1989
  Obrachevella [sic] condensata Liu, 1984
  Volyniella glomerata Jankauskas, 1980

Oscillatoriopsis Schopf, 1968
  Caudiculophycus Schopf, 1968
  Cephalophytarion Schopf, 1968
  Halythrix Schopf, 1968
  Hyalothecopsis Zhang P., 1982
  Obconicophycus Schopf & Blacic, 1971
  Partitiofilum Schopf & Blacic, 1971

Oscillatoriopsis vermiformis (Schopf, 1968) Butterfield, n.comb.
  Anabaenidium Johnsonii Schopf, 1968
  Archaeonema longicellularis Schopf, 1968
  Cephalophytarion minutum Schopf, 1968

Contortothrix vermiformis Schopf, 1968

Oscillatoriopsis obtusa Schopf, 1968 [4, 7, 16]
    Caudiculophycus acuminatus Schopf & Blacic, 1971
    Caudiculophycus rivularioides Schopf, 1968
    Cephalophytarion constrictum Schopf & Blacic, 1971
    Cephalophytarion delicatulum Schopf & Blacic, 1971
    Cephalophytarion grande Schopf, 1968
    Cephalophytarion laticellulosum Schopf & Blacic, 1971
    Cephalophytarion piliformis Mikhailova, 1986
    Cephalophytarion taenia Zhang, 1981
    Cephalophytarion turukhanicum Weiss, 1984
    Cephalophytarion variabile Schopf & Blacic, 1971
    Cyanonema disjuncta Ogurtsova & Sergeev, 1987
    Filiconstrictosus eniseicum Weiss, 1984
    Obconicophycus minor Yin, 1987
    Oscillatoriopsis anshanensis Yin, 1979
    Oscillatoriopsis breviconvexa Schopf & Blacic, 1971
    Oscillatoriopsis jixianensis Zhang, 1981
    Oscillatoriopsis luozhuangensis Zhang, 1981
    Oscillatoriopsis parvula Liu & Li, 1986
    Oscillatoriopsis psilata Maithy & Shukla, 1977
    Oscillatoriopsis qingshanensis Zhang, 1981
    Oscillatoriopsis schopfii Oehler, 1977
    Palaeolyngbya minor Schopf & Blacic, 1971
    Partitiofilum gongyloides Schopf & Blacic, 1971
    Primorivularia absoluta Hermann, 1986
    Primorivularia dissimilara Hermann, 1986

Oscillatoriopsis amadeus (Schopf & Blacic, 1971), Butterfield, n.comb. [4]
    Cephalophytarion majesticum Allison, 1989
    Obconicophycus amadeus Schopf & Blacic, 1971
    Oscillatoriopsis acuta Zhang, 1987
    Oscillatoriopsis connectens Zhang, 1987
    Oscillatoriopsis doliocellularis Zhang, 1982
    Oscillatoriopsis formosa Zhang, 1982
    Oscillatoriopsis media Mendelson & Schopf, 1982
    Oscillatoriopsis taimirica Schenfil, 1983

Oscillatoriopsis longa Timofeev & Hermann, 1979 [5, 7]
    Filiconstrictosus magnus Yakschin, 1991
    Halythrix leningradica Schenfil, 1983
    Hyalothecopsis nanshanensis Zhang, 1982
    Hyalothecopsis sinica Zhang, 1982
    Oscillatoriopsis aculeata Zhang & Yan, 1984
    Oscillatoriopsis connectens Zhang & Gu, 1986
    Oscillatoriopsis major Liu, 1982
    Oscillatoriopsis planaria Zhang & Gu, 1986
    Oscillatoriopsis princeps Zhang & Yan, 1984
    Oscillatoriopsis strictura Zhang & Gu, 1986
    Oscillatoriopsis valida Zhang & Gu, 1986
    Oscillatoriopsis variabilis Strother, Knoll & Barghoorn, 1983
    Partitiofilum tungusum Mikhailova, 1989

Osculosphaera hyalina Butterfield, n.gen., n.sp. [11*, 12]

Ostiana microcystis Hermann, 1976 [4, 5]

Palaeastrum dyptocranum n.gen., n.sp. [2, 4*, 5]

Palaeolyngbya Schopf, 1968
    Doushantuonema Zhang Z., 1981
    Rhicnonema Hofmann, 1976
    Scalariphycus Song, 1982

Palaeolyngbya catenata Hermann, 1974 [4, 21]
    Doushantuonema peatii Zhang, 1981
    Oscillatoriopsis robusta Horodyski & Donaldson, 1980
    Palaeolyngbya maxima Zhang, 1981
    Scalariphycus tianzimiaoensis Song, 1982

Palaeolyngbya hebeiensis Zhang & Yan, 1984 [4]
    Palaeolyngbya conicus [sic] Liu & Li, 1986
    Palaeolyngbya crassa Luo, 1985

Palaeolyngbya sphaerocephala Hermann & Pylina, 1986

Palaeosiphonella sp. [13]

Polybessurus bipartitus Fairchild, 1975, *ex* Green et al., 1987 [8, 9]

Proterocladus major Butterfield, n.gen., n.sp. [4*]

Proterocladus minor Butterfield, n.gen., n.sp. [4*]

Proterocladus hermannae Butterfield, n.gen., n.sp. [4*]

Pseusodendron anteridium Butterfield, n.gen., n.sp. [4, 16*]

Pseudotawuia birenifera Butterfield, n.gen., n.sp. [6*]

Pterospermopsimorpha pileiformis Timofeev, 1966 [4]

Rugosoopsis Timofeev & Hermann, 1979
    Karamia Kolosov, 1984
    Plicatidium Jankauskas, 1980
    Tubulosa Assejeva, 1982

Rugosoopsis tenuis Timofeev & Hermann, 1979 [4, 5, 14, 20]
    Karamia costata Kolosov, 1984
    Karamia jazmirii (Kolosov, 1982) Kolosov, 1984
    Karamia segmentata Kolosov, 1984
    Tubulosa corrugata Assejeva, 1982

Siphonophycus kestron Schopf, 1968 [4, 21]
    Eomycetopsis contorta Zhu, 1984
    Eomycetopsis luta Golovenoc & Belova, 1985
    Euryaulidion cylindratum Lo, 1980
    Gunflintia bruecknerii Nautiyal, 1982
    Isophyma stricta Golub, 1979
    Judomophyton unifarium Kolosov, 1982
    Omalophyma angusta Golub, 1979
    Siphonophycus beltensis Horodyski, 1980
    Siphonophycus ganjingziensis Bu, 1985
    Siphonophycus indicus Nautiyal, 1980
    Siphonophycus laishuiensis Zhang & Yan, 1984
    Siphonophycus sinensis Zhang, 1986
    Taeniatum punctatosum Du, 1985
    Uraphyton distinctum Kolosov, 1982
    Uraphyton evolutum Kolosov, 1982

Siphonophycus robustum (Schopf, 1968) Knoll et al., 1991 [4, 5, 6]
    Acranella granulata Kolosov, 1982
    Allachjunica daedalea Kolosov, 1982
    Archaeonema longicellularis Schopf, 1968
    Beckspringia communis Licari, 1978
    Eomycetopsis filiformis Schopf, 1968
    Eomycetopsis polesicus Assejeva, 1983
    Eomycetopsis psilata Maithy & Shukla, 1977
    Eomycetopsis robusta Schopf, 1968
    Eophormidium liangii Xu, 1984
    Eophormidium semicirculare Xu, 1984
    Judomophyton minisculum Kolosov, 1982
    Oscillatoriopsis acuminata Xu, 1984
    Oscillatoriopsis disciformis Xu, 1984
    Oscillatoriopsis glabra Xu, 1984
    Oscillatoriopsis hemisphaerica Xu, 1984
    Oscillatoriopsis tuberculata Xu, 1984
    Schizothropsis caudata Xu, 1984
    Siphonophycus robustum (Schopf, 1968) Knoll, 1991

Siphonophycus septatum (Schopf, 1968) Knoll et al., 1991 [4]
    Allachjunica tenuiuscula Kolosov, 1982
    Archaeotrichion lacerum Hermann, 1989
    Eomycetopsis? campylomitus Lo, 1980
    Eophormidium capitatum Xu, 1984
    Judomophyton microscopicum Kolosov, 1982
    Siphonophycus chuii Liu, 1982
    Tenuofilum septatum Schopf, 1968

Siphonophycus solidum (Golub, 1979) Butterfield, n.comb. [4]
    Eomycetopsis grandis Pjatiletov, 1988
    Leiothrichoides gracilis Pjatiletov, 1980

Omalophyma gracilis Golub, 1979
Omalophyma solida Golub, 1979
Siphonophycus capitaneum Nyberg & Schopf, 1984
Siphonophycus transvaalensis Beukes, Klein & Schopf, 1987
Solenophyma rudis Golub, 1979
Solenophyma tenuis Golub, 1979
Uraphyton crassitunicatum Kolosov, 1982
Uraphyton lenaicum Kolosov, 1982

Siphonophycus thulenema Butterfield, n.sp. [4*]

Siphonophycus typicum (Hermann, 1974) Butterfield, n.comb. [4, 6, 8, 9, 14, 18, 19, 20, 23]
Eomycetopsis crassiusculum (Horodyski, 1980) Zhang, 1982
Eomycetopsis crassus Yin, 1985b
Eomycetopsis cylindrica Maithy, 1975
Eomycetopsis pachysiphonia Zhu, 1982
Eomycetopsis pflugii Maithy & Shukla, 1977
Eomycetopsis rimata Jankauskas, 1980 *Eomycetopsis rugosa* Maithy, 1975
Eomycetopsis? siberiensis Lo, 1980
Heliconema randomensis Nautiyal, 1982
Judomophyton multum Kolosov, 1982
Judomophyton vulgatum Kolosov, 1982
Leiothrichoides tipicus Hermann, 1974
Sacharia crassa Kolosov, 1982
Siphonophycus crassiusculum Horodyski, 1980
Siphonophycus hughesii Nautiyal, 1982
Siphonophycus inornatum Zhang, 1981
Uraphyton rectum Kolosov, 1982

Sphaerophycus parvum Schopf, 1968 [15, 18, 20]

Tawuia Hofmann, 1979
Ellipsophysa Zheng, 1980
Glossophyton Duan & Du, 1985

Luonaconcha Jian & Hu, 1993
Mesonactus Fu, 1989
Nephroformia Zheng, 1980
Pumilabaxa Zheng, 1980
Tachymacrus Fu, 1989

Tawuia dalensis Hofmann, 1979 [4, 5]

Tortunema Wernadskii (Schepeleva, 1960) n.comb. [4]
Botuobia immutata Kolosov, 1984
Botuobia vermiculata Pjatiletov, 1979
Botuobia wernadskii (Schepeleva, 1960) Kolosov, 1984
Oscillatoriopsis bothnica Tynni & Donner, 1980
Oscillatoriopsis constricta Tynni & Donner, 1980
Oscillatorites Wernadskii Schepeleva, 1960
Tortunema cellulaefera Pjatiletov, 1988
Tortunema sibirica Hermann, 1976

Trachyhystrichosphaera aimika Hermann, 1976 [4, 16, 18, 23]
Nucellohystrichosphaera megalea Timofeev & Hermann, 1976
Nucellosphaeridium bellum Timofeev, 1969
Trachyhystrichosphaera cyathophora Hermann, 1989
Trachyhystrichosphaera magna Allison, 1989
Trachyhystrichosphaera megalia (Timofeev, 1976) Pjatiletov, 1988
Trachyhystrichosphaera membranacea Pjatiletov, 1988
Trachyhystrichosphaera stricta Hermann, 1989
Trachyhystrichosphaera vidalii Knoll, 1984

Trachyhystrichosphaera polaris Butterfield, n.sp. [4*]

Valkyria borealis Butterfield, n.gen., n.sp. [4*]

Veteronostocale Schopf & Blacic, 1971
Filiconstrictosus Schopf & Blacic, 1971

Veteronostocale amoenum Schopf & Blacic, 1971 [4]
Filiconstrictosus diminutus Schopf & Blacic, 1971